Anderson Acceleration for Numerical PDEs

SIAM Spotlights

SIAM Spotlights is a new book series that comprises brief and enlightening books on timely topics in applied and computational mathematics and scientific computing. The books, spanning 125 pages or less, will be produced on an accelerated schedule and will be attractively priced.

Josef Málek and Zdeněk Strakoš, *Preconditioning and the Conjugate Gradient Method in the Context of Solving PDEs*

Paul G. Constantine, *Active Subspaces: Emerging Ideas for Dimension Reduction in Parameter Studies*

Dominique Orban and Mario Arioli, *Iterative Solution of Symmetric Quasi-Definite Linear Systems*

Lin Lin and Jianfeng Lu, *A Mathematical Introduction to Electronic Structure Theory*

A. Målqvist and D. Peterseim, *Numerical Homogenization by Localized Orthogonal Decomposition*

Gérard Meurant and Petr Tichý, *Error Norm Estimation in the Conjugate Gradient Algorithm*

Daan Camps, Thomas Mach, Raf Vandebril, and David S. Watkins, *Pole-Swapping Algorithms for the Eigenvalue Problem*

Sara Pollock and Leo G. Rebholz, *Anderson Acceleration for Numerical PDEs*

Anderson Acceleration for Numerical PDEs

Sara Pollock
University of Florida, Gainesville, Florida

Leo G. Rebholz
Clemson University, Clemson, South Carolina

Society for Industrial and Applied Mathematics
Philadelphia

10 9 8 7 6 5 4 3 2 1

Publications Director	Kivmars H. Bowling
Executive Editor	Elizabeth Greenspan
Acquisitions Editor	Paula M. Callaghan
Developmental Editor	Rose Kolassiba
Managing Editor	Kelly Thomas
Production Editor	David Riegelhaupt
Copy Editor	David Riegelhaupt
Production Manager	Rachel Ginder
Production Coordinator	Cally A. Shrader
Compositor	Cheryl Hufnagle
Graphic Designer	Doug Smock

Library of Congress Control Number: 2025940228

Contents

Preface

Nonlinear solvers are a fundamental technology in computational mathematics: solving finite-dimensional systems of equations is the key step in obtaining a solution to a descriptive or a predictive model. Anderson acceleration (AA) is a low-cost technique used to improve the efficiency and robustness of solvers for discrete nonlinear (and linear) systems across a wide array of applications in computational mathematics. It uses a history of a sequence of approximate solutions and correction steps to improve the next approximation in a sequence.

There has been an explosion of research on AA in the last 10-15 years, including fundamental results such as the establishment of convergence theory, implementation improvements such as filtering techniques, advances in parameter selection methods, and application to a wide variety of problems across much of science and engineering. While the technique was introduced over 50 years ago, the theoretical understanding of AA has largely been developed more recently. Several of the big questions about the method have been answered; for instance, isolating the mechanism by which the method accelerates standard iterative methods. However, many questions remain, including specifics of optimal implementations for various types of applications.

The focus of this book will be the application of AA to partial differential equations (PDEs). PDEs are ubiquitous across mathematics, science, engineering, and economics as predictive models for a vast quantity of important phenomena, from fluid and solid mechanics to combustion to stock market forecasting to (perhaps?) flux capacitors. We will necessarily exclude results on AA specific to applications in optimization and machine learning algorithms; these are interesting topics in and of themselves, but fall outside the scope of the current presentation.

We (the authors) have been working on and with AA since a 2015 ICERM workshop attended by author SP. SP was kind enough to bring in author LR in 2018, after she saw a possible connection of AA to a solver presented at the 2018 JMM by Mengying Xiao, a Ph.D. student of LR at the time (and later a postdoc of SP, and now tenure track at the University of West Florida). Since then, between us and our students we have written 15 papers about AA theory and applications, many under a joint National Science Foundation grant, Advancing Theoretical Understanding of Accelerated Nonlinear Solvers, with Applications to Fluids (NSF DMS 2011490/2011519),[1] solving some (in our view) important mathematical and algorithmic problems along the way. We believe we have reached a point where it makes sense to compile our results into a book. If there was any doubt, author LR recalled once attending a workshop on book writing at a SIAM annual meeting where panelist and well-known applied mathematics textbook author Tim Kelley stated that "*once you reach 100 pages, you have a book.*" ; we have reached the threshold.[2]

The aim of the book is threefold: First, we aim to collect recent fundamental results for AA, both theory and application. It will be helpful to both researchers and practitioners if the latest and best results are compiled together in an easy to find location. Second, we want the book

[1]Author LR also acknowledges support from Department of Energy grant DE-SC0025292.

[2]Three conferences are mentioned in this paragraph. Authors SP and LR both credit a significant amount of their career successes to traveling to important conferences such as SIAM, AMS, Finite Element Circus, JMM, and talking (and listening!) to other mathematicians there.

to be useful as a textbook for graduate students in applied mathematics, scientific computing, and computer science. We will give the main theoretical results with proofs, and most of the proofs are within reach for a 2nd year graduate student who has had a 2-course sequence in numerical analysis / scientific computing. Third, the book will be a resource for practitioners. We provide results and implementation details for many test problems herein, using a wide range of PDEs. Our testbed includes Navier-Stokes, Boussinesq, p-Laplace, Gross-Pitaevskii, nonlinear Helmholtz, and Bingham models. For these tests, we give codes and show how we use AA (e.g., parameter choices) to get good results. While we of course cannot solve every nonlinear PDE herein, from what we do provide we expect that the ideas can be extrapolated to many other interesting problems.

Acknowledgments

We acknowledge Matt Dallas, Duygu Vargun, Mengying Xiao, and Fei Xue for their research collaborations and coauthorship on many of the results and papers that led to this book. We also acknowledge Matt Dallas and Jessica Franklin for their comments on an earlier version of the book.

Chapter 1

Introduction

Before you criticize someone, you should walk a mile in their shoes. That way when you criticize them, you are a mile away from them and you have their shoes.
—Jack Handey

Nonlinear problems are ubiquitous throughout physical modeling, mathematics and beyond. Anderson acceleration (AA) has rapidly become an enabling technology for the solution of nonlinear problems in scientific computing and computational mathematics. It has been shown in hundreds of works over the last several years to improve the effectiveness of nonlinear solvers by improving their convergence properties, both by accelerating convergence (achieving the tolerance in fewer iterations) and yielding convergence when the solver would otherwise fail. An example of the latter relevant to this book is the application of AA to the Picard iteration for the Navier-Stokes equations, which allows their solution at higher Reynolds numbers than was previously possible [88, 91]).

The method was initially developed in the mid 1960s by D.G. Anderson [6] and became heavily used since then in certain areas such as computational chemistry. After the 2011 paper of Walker and Ni [123], there was an explosion of interest in AA by the computational science community at large. This paper showed how AA can be used to get substantially better convergence properties on a variety of problems for essentially no additional computational cost. While of course this was true since AA was first introduced, this paper was the spark that got many researchers to *try* AA; and this was enough to get many (if not most) to realize its tremendous power and to begin using it as a matter of routine.

While AA now enjoys widespread use, developments in theory for the method have been more limited. A solid grasp of the theory for a mathematical algorithm leads to better understanding of when it works, how to improve it, and how to pick parameters.[3] In terms of connections to known methods with solid theoretical backing, AA was analyzed in [52, 53] as a generalized multi-secant or quasi-Newton algorithm, and also discussed within a Krylov space framework in [123]. In [27], a connection between AA and classical extrapolation algorithms used to accelerate vector sequences, including the (Modified) Minimal Polynomial Extrapolation, Topological and Vector Epsilon algorithms, and Reduced Rank Extrapolation algorithms was established. Local convergence of the iteration was first shown in [120] and then sharpened in [32, 77], and here it

[3]Most computational scientists acknowledge that improvement in algorithms over the last 50 years has contributed just as much to computational improvements than the improvements in processor speed. Solid theoretical analysis of algorithms is a critical aspect to this.

was shown that AA "can do no harm," asymptotically when applied to contractive operators. The acceleration property of the method was first theoretically developed in [50, 88, 91], and this will be discussed in detail in this book. Analytical results that partially address the question of AA asymptotic linear convergence for the general case are found in [114, 115, 116] and for application to the Alternating Direction Method of Multipliers (ADMM) in [124]. Very recently there are also some analytical results for the case of nonsmooth fixed point problems [17, 18]. Some convergence results for a stabilized version of AA are given in [25], and these authors also give some new interpretations of AA. More details on some of these topics are given in Chapter 2.

The popularity of AA is due to its effectiveness and its ease of use. It has been used to improve solvers in various flow problems [21, 50, 81, 91], geometry optimization [84], electronic structure computations [52, 53] (closely related methods for this application are mentioned in the next chapter), radiation diffusion and nuclear physics in [5, 121], computing nearest correlation matrices in [73], accelerating Douglas-Rachford splitting convergence [59], and molecular interaction in [112]. This is a very abbreviated list, and if one scrolls through the titles of the nearly 700 papers that cite the Walker and Ni paper, it is hard to find a hard nonlinear problem where AA is not being used.

1.1 ▪ Anderson acceleration's role in the solution process

We now discuss the PDE solution process and how AA fits into it. In the context of approximating solutions to PDEs, one typically begins with a system of the form

$$F(u) = 0,$$

together with boundary conditions. Here u is the solution and F contains all the necessary information for the differential system, e.g., $F(u) = L(a)u + N(a,u) - f(a)$, where $L(a)$ and $N(a,u)$ are the linear and nonlinear differential operators with respect to the solution u, and a denotes the dependence on any parameters or physical coordinates. Several examples of nonlinear PDE systems that we will use as examples throughout this book are given in Chapter 3. These examples are chosen to highlight the use of AA as an enabling technology for different classes of nonlinear PDEs. We will see these different problems characterize different challenges in the solution process. We generally present the steady versions of these problems; however, the techniques presented could also be used to determine the solution at each time step in the time-dependent case. Of course, whether a solution exists and whether it is unique or locally unique are important considerations that should not be ignored.

The next step in the solution process is discretizing the nonlinear problem. In order to turn an infinite dimensional problem into a finite dimensional system that can be solved on a computer, approximations must be made to things such as how derivatives are represented (finite differences and certain generalizations of finite elements) or subspaces over which the solution satisfies the PDE (finite elements and generalizations thereof). Assuming the boundary conditions can be built into the discretization, which is often the case with finite elements or finite differences, the nonlinear system one wishes to solve can be represented by

$$F_h(u_h) = 0.$$

Solution existence and uniqueness must again be considered, as well as the approximation error $\|u - u_h\|$ (and in the appropriate norm), before moving ahead to try to solve this nonlinear system.

With the finite dimensional nonlinear PDE defined, one then faces the task of solving the nonlinear system. Here, one will typically attempt to devise an iteration to solve $F_h(u_h) = 0$. Often a first attempt is Newton's method - *if Newton's method works and is efficient for any particular problem, use it!* Unfortunately, Newton's method has limitations when it comes to nonlinear PDEs. A common issue is that for PDEs that get harder with more complex solutions as certain parameters get bigger (or smaller), Newton becomes less effective and eventually fails for big (small) enough parameters. This happens, for example, in fluid mechanics as Reynolds numbers or Rayleigh numbers get larger, and the values that cause Newton to fail are still well within the span of relevant physical settings. A discussion of improving the convergence of Newton's method with AA at or near bifurcation points is included in Chapter 7 as introduced in [37, 38]. Another problem that can occur with Newton's method is that its implementation can be difficult; for example, in coupled nonlinear systems the Newton linearization will generally produce fully coupled linear systems that need to be solved at each iteration. And finally, the success of Newton iterations is in most cases tied to providing a sufficiently good initial iterate. Hence there are many times when Newton's method (alone) will not be the best solver.

Once it is established that Newton's method is not going to efficiently solve your nonlinear system, then the process of devising a good fixed point iteration begins. Good is of course a relative term, but desirable properties are that it is efficient (reasonable linear system solves at each iteration), stable (solutions at each iteration remain bounded), and convergent (perhaps contractive).

To reduce doubly subscripted notation, here and throughout the book we will denote our sequence of iterates that approximate the solution to $F_h(u_h) = 0$ by $\{x_k\}_{k\geq 0}$. That is, one wishes to define / construct a function g so that the following hold:

1. The solution x^* to $g(x) = x$ satisfies $F_h(x^*) = 0$, where without confusion x^* is associated with u_h.

2. It is easy to evaluate $g(x)$ for x near to x^*.

3. $\|g(x)\|$ is bounded in the appropriate norm for all x (or at least x near x^*).

4. The sequence defined by $x_{k+1} = g(x_k)$ converges to x^*.

Constructing such a g is often difficult, and proving that your g has these properties (if it does) can also be challenging. Several examples of solution operators g that correspond to nonlinear problems $F(u) = 0$ in the context of finite element approximations are given in Chapter 3.

We finally arrive at the application of AA, which works by improving the convergence of fixed point iterations

$$x_{k+1} = g(x_k),$$

where a sequence of iterates $\{x_k\}$ are approximations to a fixed point of (nonlinear) operator g. AA involves a correction to each fixed-point update that is formed from a linear combination of a history of iterates and update steps, where the coefficients of the linear combination used in the update are recomputed at each iteration by the solution to an optimization problem which determines a least-length update step. For many iterations and many applications, AA has been shown to improve and even enable the convergence of fixed-point iterations. The optimization step typically has negligible computational cost compared to the linear solve that gets done at each iteration. The main requirement over a standard fixed-point iteration is that some number of older solution and update vectors must be stored. The next subsection presents the AA algorithm.

1.2 ▪ The Anderson acceleration algorithm

AA seeks a fixed-point x^* of $g(x)$ via the fixed-point iteration $x_{k+1} = g(x_k)$, where $g : X \to X$ for a Hilbert space X. The Hilbert space X will be equipped with norm $\|\cdot\|$ and inner product $(\cdot,\cdot)$. As we are concerned with the case of discretized PDEs, we will assume $x \in X_h \subset X$, where X_h is finite dimensional. Assumptions on smoothness will be made about g later in this book as we develop theory to characterize convergence and acceleration.

While the AA algorithm in its most general form does not identify the norm that should be used in the optimization, it is common and efficient to consider the minimization in the l^2 sense. Then we can solve the least-squares problem that arises at each step with a (fast) QR [120, 123]. However, our analysis uses the X norm and inner product (e.g., $X = H_0^1$ for Navier-Stokes) for our theoretical results that show AA improves the linear convergence rate. Here we include a discussion in Section 1.2.2 on using a scaled l^2 norm (and inner product) in the computation to allow the implementation to agree with the theory.

Writing our fixed-point iteration as $x_{k+1} = g(x_k)$ in terms of an update step $x_{k+1} = x_k + w_{k+1}$, it is natural to define the difference between iterates e_k and the residual (or update step) w_k by

$$e_k := x_k - x_{k-1}, \qquad w_{k+1} := g(x_k) - x_k. \tag{1.2.1}$$

Then AA can be written as follows, where parameter m is the maximum allowable algorithmic depth, m_k is the algorithmic depth at iteration k, and β_k is the relaxation (also called damping or mixing) parameter used on iteration k.

ALGORITHM 1.1. Anderson acceleration.

The algorithm starts with a fixed-point update step ($k = 0$); then the acceleration starts at $k = 1$.

Choose initial iterate x_0 and algorithmic depth parameter m

Compute w_1, set β_0, and update $x_1 = x_0 + \beta_0 w_1$ $\quad\triangleright\ k = 0$

1: **for** $k = 1, \ldots$ **do** $\quad\triangleright\ k > 0$

2: Compute w_{k+1}

3: Set $m_k = \min\{k, m\}$

4: Set $F_k = \big((w_{k+1} - w_k) \ \ \ldots \ \ (w_{k+1-(m_k-1)} - w_{k-(m_k-1)})\big)$

5: and $E_k = \big((e_k) \ \ \ldots \ \ (e_{k-(m_k-1)})\big)$

6: Find $\gamma^{k+1} = \operatorname{argmin} \|F_k\gamma - w_{k+1}\|$

7: Set relaxation parameter $\beta_k \in (0, 1]$

8: Update $x_{k+1} = x_k + \beta_k w_{k+1} - (E_k + \beta_k F_k)\gamma^{k+1}$

9: **end for**

An equivalent [53] AA algorithm that uses a constrained minimization is given next. The algorithm above is used for the analysis of the method in [88, 89], and better suited if adding filtering to improve convergence and control the condition / enforce sufficient linear independence of the history of udpates used, as we present in Chapters 4 and 5. Without such control, assumptions or arrangements are necessary to control the magnitude of the optimization coefficients γ^{k+1} from step 6. The algorithm below was used for analysis of the method [77, 91, 120] and would allow for constrained optimization (keeping the α_j^{k+1} coefficients in [0,1], for example). Either algorithm can accommodate rescaling the l^2 norm to work in spaces such as $X \subseteq H_0^1$. Sometimes the software one is using can dictate which of these implementations is easier to use. There is a complete lack of agreement in the community at large (and indeed among the current authors) about which version is more natural or intuitive.

ALGORITHM 1.2. Anderson iteration in constrained optimization form.

The algorithm starts with a fixed-point update step ($k = 0$); then the acceleration starts at $k = 1$.

Choose initial iterate x_0 and algorithmic depth parameter m

Compute w_1, set β_0, and update $x_1 = x_0 + \beta_0 w_1$ $\quad\triangleright\ k = 0$

1: **for** $k = 1, \ldots$ **do** $\quad\triangleright\ k > 0$

2: Compute w_{k+1}
3: Set $m_k = \min\{k, m\}$
4: Solve the minimization problem for $\{\alpha_j^{k+1}\}_{k-m_k}^k$

$$\min_{\sum_{j=k-m_k}^k \alpha_j^{k+1}=1} \left\| \sum_{j=k-m_k}^{k} \alpha_j^{k+1} w_{j+1} \right\| \tag{1.2.2}$$

5: Set relaxation parameter $\beta_k \in (0, 1]$
6: Update $x_{k+1} = \sum_{j=k-m_k}^k \alpha_j^{k+1} x_j + \beta_k \sum_{j=k-m_k}^k \alpha_j^{k+1} w_{j+1}$
7: **end for**

1.2.1 ▪ Relating Algorithms 1.1 and 1.2

The updates of line 8 in Algorithm 1.1 and line 6 in Algorithm 1.2 are (in exact arithmetic) the same and can be written in terms of the averages x_k^α and w_{k+1}^α by

$$x_{k+1} = x_k^\alpha + \beta_k w_{k+1}^\alpha. \tag{1.2.3}$$

In terms of Algorithm 1.1, the averages x_k^α and w_{k+1}^α are given by

$$x_k^\alpha = x_k - E_k \gamma^{k+1}, \quad w_{k+1}^\alpha = w_{k+1} - F_k \gamma^{k+1}. \tag{1.2.4}$$

In terms of Algorithm 1.2, they are given by

$$x_k^\alpha = \sum_{j=k-m_k}^{k} \alpha_j^{k+1} x_j, \quad w_{k+1}^\alpha = \sum_{j=k-m_k}^{k} \alpha_j^{k+1} w_{j+1}. \tag{1.2.5}$$

The sets of optimization coefficients in (1.2.4) and (1.2.5) are related by

$$\gamma_j^{k+1} = \sum_{n=k-m_k}^{j-1} \alpha_n^{k+1}, \quad \text{where } \gamma^{k+1} = (\gamma_k^{k+1}, \gamma_{k-1}^{k+1}, \ldots, \gamma_{k-m_k+1}^{k+1})^T. \tag{1.2.6}$$

By this transformation, it is a straightforward exercise to see the optimization problem stated in Algorithm 1.2 as (1.2.2), is equivalent to the unconstrained minimization problem $\gamma^{k+1} = \operatorname{argmin}_{\gamma \in \mathbb{R}^{m_k}} \|w_{k+1} - F_k \gamma\|$ that appears in Algorithm 1.1.

We next define the stage-k gain θ_k which quantifies the success of the optimization by

$$\|w_k^\alpha\| = \theta_k \|w_k\|. \tag{1.2.7}$$

As we will see in the chapters that follow, this important quantity scales down the first-order term in the residual expansion. In the original fixed-point iteration (before acceleration), that first-order term is the entire residual bound. Along with scaling down the first-order term, AA also introduces additional higher-order terms to the residual bound. These higher-order terms will be shown in Chapter 4 to be scaled by a factor of $\sqrt{1-\theta_k^2}$, meaning a successful optimization increases the relative weight of the higher-order terms, and an unsuccessful optimization increases the relative weight of the first-order term in the residual expansion.

1.2.2 ▪ Rescaling the optimization problems

One of the advantages of using Algorithm 1.2 is the ease in which a rescaling of the l^2 norm can be applied. An efficient way to implement the minimization problem in line 4 is as follows. For simplicity, fix $k > m$ so that we can focus on a single iteration and write

$$W = [w_{k+1}\; w_k\; w_{k-1}\; \dots\; w_{k-m+1}], \text{ and } \alpha = [\alpha_k^{k+1}\; \alpha_{k-1}^{k+1}\; \dots\; \alpha_{k-m}^{k+1}].$$

Now the minimization problem at step k can be equivalently written as

$$\operatorname{argmin}_{\sum \alpha_i = 1} \|W\alpha\|_X^2, \text{ where } \|W\alpha\|_X^2 = \alpha^T W^T M W \alpha.$$

Performing a Cholesky factorization to get $R^T R = W^T M W$, we have

$$\|W\alpha\|_X^2 = \alpha^T (R^T R)\alpha = \|R\alpha\|^2,$$

where the norm on the right is an l^2 norm. For most common discretization methods, the computation of $W^T M W$ will not be expensive as the matrix M is sparse. At each iteration one can reuse information from the prior iteration (since W at each iteration shifts columns by 1 to the right and adds a new column), and the number of columns of W is small since m is typically small (often less than 10). Thus the minimization problem has now been shown as equivalent to solving the $(m+1) \times (m+1)$ linear system

$$R\alpha = 0$$

subject to the constraint that $\sum \alpha_i = 1$. Denoting the columns of R by r_i and inserting the constraint into the linear system we obtain

$$\alpha_1 r_1 + \alpha_2 r_2 + \cdots + \alpha_m r_m + (1 - \alpha_1 - \alpha_2 - \cdots - \alpha_m) r_{m+1} = 0,$$

which can be rearranged as

$$\alpha_1(r_1 - r_{m+1}) + \alpha_2(r_2 - r_{m+1}) + \cdots + \alpha_m(r_m - r_{m+1}) = -r_{m+1}.$$

Constructing the $(m+1) \times m$ matrix

$$A = [(r_1 - r_{m+1})\; (r_2 - r_{m+1})\; \dots\; (r_m - r_{m+1})],$$

we can now write the constrained minimization problem as the $(m+1) \times m$ linear least squares problem

$$A\alpha = -r_{m+1},$$

which can be easily solved and which enforces the constraint $\sum_i \alpha_i = 1$.

1.2.3 ▪ Theoretical understanding and implementation

Donald Anderson, the originator of AA, revisited his method (which he called "The Extrapolation Method") in 2019 in [7], appreciating the breadth of applications and scale of modern problems it was being used for. One of the key points he mentioned is the method is not meant to be stationary. The damping (relaxation) factors β_k may change at each iteration, the algorithmic depth m_k may change, or different columns from the history may be selected, as the iteration progresses.

In the presentation that follows, particularly in Chapters 4 and 5, we present dynamic parameter updates and filtering strategies based on our theoretical analysis of the method. Our approach is through analyzing the residual expansion (assuming sufficient smoothness of the underlying fixed-point operator). We then choose parameters to increase the depth when the additional contributed higher-order terms will do no harm, and to filter the iterates and update steps used to maintain the sufficient linear independence between columns of the matrix used in the least squares problems, and ultimately to ensure it is well conditioned. We give some examples and a heuristic for choosing the dynamic damping parameters β_k to enable convergence in the preasymptotic regime where increasing the depth would be detrimental to convergence; however, the development of robust dynamic damping strategies remains an open problem as of this writing.

Chapter 2

Theoretical context

Whatever you do, always give 100%. Unless you are donating blood.

—Bill Murray

The theory and ideas presented in the remaining chapters of this book were developed based on the insights generated by mathematicians, computational scientists (especially computational chemists), and nonlinear solver folks who worked on these problems before we did. In this chapter we aim to give a concise summary of those ideas and insights, and to point out some of the connections, ideas, and advances that have been developed based on them.

A great deal of work on AA and related methods has been done in the context of electronic structure calculations, e.g., in conjunction with a self-consistent field (SCF) iteration [52, 53, 120, 123]. Very closely related methods include Pulay mixing [96, 97] and direct inversion on the iterative subspace (DIIS) [102, 106]. Some explicit connections between these methods can be found in [34]. The work discussed below in sections 2.1-2.2 all emerged either directly or indirectly from work in that application area. SCF iterations are sufficiently complicated that an analytic Jacobian can be considered unavailable, each function evaluation is expensive, and the dimension of the problem is high. These features make a Jacobian-free post-processing procedure such as AA an attractive technique, and even an enabling technology if the technique is well enough understood to be implemented efficiently and robustly.

As explained nicely in [53], the SCF iteration is also the source of the term *Anderson mixing*, another name by which AA is sometimes known. In this iteration both the input to and the output of the fixed-point iteration is a potential V. The term historically used in the electronic structure literature that creates a new approximate potential V_{k+1} from a sequence of approximations $V_{k-m}, \ldots V_k$ is *mixing*. Hence in this context, AA is naturally referred to as Anderson *mixing*.

2.1 ▪ Connections to quasi-Newton methods

The characterization of AA as an approximate Newton or *quasi-Newton* method, and specifically as a Type-II generalized Broyden method, was introduced by Eyert in [52], where a useful formulation of a generalized Broyden method is also introduced. A further extension of this analogy leading to the introduction of a Type-I AA method (among other results) was presented by Fang and Saad in [53]. Here we summarize the main results along those lines following the discussion in [53]. A related discussion can be found in [123].

Table 2.1. *A comparison between the notation used here and the notation used in [53].*

Notation	$f(x)$	$x_{k+1} - x_k$	$f(x_k)$	$f(x_{k+1}) - f(x_k)$
[53]	$f(x)$	Δx_k	f_k	Δf_k
This book	$g(x) - x$	e_{k+1}	w_{k+1}	δw_{k+2}

Broyden's methods [28] for solving the system of equations $f(x) = 0$ are computationally efficient approximations to Newton's method, $x_{k+1} = x_k - f'(x_k)^{-1} f(x_k)$, that construct an approximate Jacobian $J_k \approx f'(x_k)$, or an approximate inverse Jacobian $G_k \approx f'(x_k)^{-1}$, at each iteration, by a rank-1 update of the approximate Jacobian (or inverse Jacobian) used at the previous iteration. The approximate Jacobians are designed to enforce two key properties: a secant condition and a no-change condition, as discussed below. To lend some clarity to the differences in notation used in this book and in [53], we include Table 2.1 to relate the key quantities used in the remainder of the discussion, starting with the mapping between the zero-finding problem $f(x) = 0$, with the fixed-point problem $g(x) = x$, via $f(x) = g(x) - x$.

2.1.1 ▪ Broyden's methods

Broyden's method of Type I, known as Broyden's first method, defines a sequence of approximate Jacobians $J_k \approx J(x_k)$, where x_k is the current approximation to the solution x of $f(x) = 0$. The Jacobian update is designed to satisfy two properties, the first of which is the secant condition, $J_{k+1}(x_{k+1} - x_k) = f(x_{k+1}) - f(x_k)$. For $f : \mathbb{R}^n \to \mathbb{R}^n$, the $n \times n$ exact Jacobian satisfies n secant conditions, which leads to generalized Broyden methods as discussed below. The original Broyden method requires the "latest" secant condition to hold. Following the notation in Table 2.1 we can write this as

$$J_k e_k = \delta w_{k+1}. \tag{2.1.1}$$

The second enforced property is the no-change condition

$$J_k q = J_{k-1} q \text{ for all } q \text{ such that } q^T e_k = 0. \tag{2.1.2}$$

This means the new approximate Jacobian does not introduce information from directions which we do not have available information. These two properties lead to a unique [28] Jacobian update

$$J_k = J_{k-1} + (\delta w_{k+1} - J_{k-1} e_k)(e_k^T e_k)^{-1} e_k^T. \tag{2.1.3}$$

Broyden's method of Type II, known as Broyden's second method, defines instead an update to an approximation of the inverse Jacobian. It is also defined to satisfy a secant condition and a no-change condition. The secant condition for the Type-II method is

$$G_k \delta w_{k+1} = e_k, \tag{2.1.4}$$

and the no-change condition is

$$G_k q = G_{k-1} q \text{ for all } q \text{ such that } q^T e_k = 0. \tag{2.1.5}$$

The requirements (2.1.4) and (2.1.5) lead to the unique update

$$G_k = G_{k-1} + (\delta e_k - G_{k-1} \delta w_{k+1})(\delta w_{k+1}^T \delta w_{k+1})^{-1} \delta w_{k+1}^T. \tag{2.1.6}$$

Finally, applying the Sherman-Morrison-Woodbury formula [108, 126] for the inverse of a low-rank update, the Type-I update (2.1.3) can be written in terms of G_k as

$$G_k = G_{k-1} + (\delta e_k - G_{k-1} \delta w_{k+1})(e_k^T G_{k-1} \delta w_{k+1})^{-1} e_k^T G_{k-1}. \tag{2.1.7}$$

Table 2.2. *A comparison between the matrix notation used here and the matrix notation used in [53]. By referring to Table 2.1, we can see the corresponding matrices contain the same vectors as columns, but in the opposite order.*

[53]	$\mathscr{F}_k = \begin{pmatrix}\Delta f_{k-m} & \dots & \Delta f_{k-1}\end{pmatrix}$	$\mathscr{X}_k = \begin{pmatrix}\Delta x_{k-m} & \dots & \Delta x_{k-1}\end{pmatrix}$
This book	$F_k = \begin{pmatrix}\delta w_{k+1} & \dots & \delta w_{k-m+2}\end{pmatrix}$	$E_k = \begin{pmatrix}e_k & \dots & e_{k-m+1}\end{pmatrix}$

This formulation prevents the need to compute the inverse of G_k at each iteration of the algorithm. Plugging either (2.1.7) or (2.1.6) into the approximate Newton update for the inverse Jacobian $x_{k+1} = x_k - G_k w_{k+1}$ yields the Type-I and Type-II Broyden iterations.

Another important way to characterize (or derive) the Broyden updates is shown in [44]: the Type-I Jacobian update for J_k minimizes $\|J_k - J_{k-1}\|_F^2$ subject to the secant condition (2.1.1), where $\|\cdot\|_F$ denotes the Frobenius norm. Similarly, the Type-II inverse Jacobian update for G_k minimizes $\|G_k - G_{k-1}\|_F^2$ subject to the secant condition (2.1.4).

2.1.2 ▪ Generalizing Broyden's methods

In [52] it was shown that Broyden's methods of Type I/II can be generalized from a rank-1 (inverse) Jacobian update satisfying a single secant condition to a rank-m (inverse) Jacobian update satisfying m secant conditions. For the Type-II update, written for the approximate inverse Jacobian G_k, we can write m consecutive secant conditions as

$$G_k \delta w_{i+1} = e_i, \quad i = k-m+1, \dots, k. \tag{2.1.8}$$

Referring to the matrix notation Table 2.2, we can write (2.1.8) in matrix form as

$$G_k F_k = E_k. \tag{2.1.9}$$

The m consecutive no-change conditions for the approximate inverse Jacobian are

$$(G_k - G_{k-m})q = 0 \text{ for all } q \in \text{null}(F_k^T), \tag{2.1.10}$$

by which the range of $(G_k - G_{k-m})^T$ is precisely the range of F_k, or equivalently the range of $(G_k - G_{k-m})$ is precisely the range of F_k^T. Thus $G_k - G_{k-m} = ZF_k^T$ for a matrix Z, which can be found explicitly by multiplying both sides of the last equation by F_k, assuming that F_k has full column rank. This yields $Z = (G_k - G_{k-m})F_k(F_k^T F_k)^{-1}$. Applying (2.1.9) allows

$$\begin{aligned} G_k &= G_{k-m} + (G_k - G_{k-m})F_k(F_k^T F_k)^{-1}F_k^T \\ &= G_{k-m} + (E_k - G_{k-m}F_k)(F_k^T F_k)^{-1}F_k^T, \end{aligned} \tag{2.1.11}$$

an update for the approximate inverse Jacobian.

Applying (2.1.11) to the approximate Newton update $x_{k+1} = x_k - G_k w_{k+1}$ yields the Type-II generalized Broyden method

$$\begin{aligned} x_{k+1} &= x_k - G_{k-m}w_{k+1} + (E_k - G_{k-m}F_k)(F_k^T F_k)^{-1}F_k^T w_{k+1} \\ &= x_k - G_{k-m}w_{k+1} + (E_k - G_{k-m}F_k)\gamma^{k+1}, \end{aligned} \tag{2.1.12}$$

where $\gamma^{k+1} = \text{argmin}\,\|F_k\gamma - w_{k+1}\|$, as in Algorithm 1.1, formulated via the normal equations (which we do not recommend using computationally). Notice that replacing G_{k-m} in (2.1.11) with $-\beta I$ yields the update step of line 8 in AA Algorithm 1.1.

As discussed in [53, Section 3.1], we can follow the procedure above with the secant equations and no-change conditions written for the approximate Jacobian update J_k to determine a Type-I generalized Broyden method. This yields the update for J_k as

$$J_k = J_{k-m} + (F_k - J_{k-m}E_k)(E_k E_k^T)^{-1}E_k^T,$$

or applying the Sherman-Morrison-Woodbury formula to write the update in terms of G_k

$$G_k = G_{k-m} + (E_k - G_{k-m}F_k)(E_k^T G_{k-m}F_k)^{-1}E_k^T G_{k-m}. \quad (2.1.13)$$

As discussed in [53], both (2.1.11) and (2.1.13) can be written in the form

$$G_k = G_{k-m} + (E_k - G_{k-m}F_k)V_k^T, \quad (2.1.14)$$

where $V_k^T F_k = I$ enforces the secant condition $G_k F_k = E_k$, and the updates (2.1.11) and (2.1.13) are two instances of a larger class defined by of all such choices of V_k^T.

The update for x_{k+1} corresponding to (2.1.14) is

$$x_{k+1} = x_k - G_{k-m}w_{k+1} - (E_k - G_{k-m}F_k)V_k^T w_{k+1}. \quad (2.1.15)$$

For the Type-II update, for which $V_k^T = (F_k^T F_k)^{-1}F_k$, we noticed previously that substituting $-\beta I$ for G_{k-m} yields the AA update of Algorithm 1.1. This update is identified as existing in the class of Type-II Broyden methods. Similarly, substituting $-\beta I$ for G_{k-m} in the Type-I update, for which $V_k^T = (E_k^T G_{k-m}F_k)^{-1}E_k^T G_{k-m}$, from (2.1.13), yields what is introduced in [53] as Type-I AA. This update is given explicitly for x_{k+1} as

$$x_{k+1} = x_k + \beta w_{k+1} - (E_k + \beta F_k)(E_k^T F_k)^{-1}E_k^T w_{k+1}. \quad (2.1.16)$$

We will point out some recent innovations in the understanding and applications of this iteration in Section 2.4.2.

Similarly to the rank-1 update Broyden methods, the rank-m methods given by the updates (2.1.11) and (2.1.13) are the unique methods to satisfy their corresponding sets of secant equations and no-change equations. They are also characterized by minimizing $\|J_k - J_{k-m}\|_F^2$ (Type-I) and $\|G_k - G_{k-m}\|_F^2$ (Type-II) subject to the corresponding set of secant conditions. In this sense, the Anderson Type-II (usual AA) update implicitly constructs an "approximate inverse Jacobian" G_k that minimizes $\|G_k + \beta I\|_F^2$ subject to the secant conditions (2.1.9), and the analogous statement can be made for the Type-I method (2.1.16).

In the discussion above, we assumed that matrix F_k maintains full column rank. We will return in Chapter 5 to techniques that enforce the condition of the least-squares problem solved in Type-II methods. We will refer there to a discussion included in [53] about using a truncated singular value decomposition (SVD) or rank revealing QR factorization as standard approaches.

Remark 2.1. *It is worth mentioning again here the different column ordering used in both [53] and [123], in contrast to [88, 89] and this book. Specifically, as noted in Table 2.2 the columns in* E_k, F_k *(this work) are ordered from left to right as most recent to oldest. The columns of* $\mathscr{X}_k, \mathscr{F}_k$ *are ordered from left to right oldest to most recent. Each respective ordering agrees with the recommended techniques for updating and maintaining the condition of the least-squares problem of Algorithm 1.1. In particular, in [123], at each iteration a new column on the right is added and older columns (on the left) are consecutively dropped. The least-squares problem is solved using a QR decomposition which is efficiently updated using Givens rotations [65, Section 12.5.2]. The ordering of [53, 123] is the most efficient for these updates. On the other hand, in [88, 89] and here, a QR decomposition is also used (both analytically and numerically) for Algorithm 1.1,* *but the column filtering strategies can remove any column but the first (as determined by the diagonal values of R in the QR decomposition) as well as dropping older columns to maintain condition of the matrix. It is natural in this analysis to keep the newest information on the left as the QR decomposition is naturally analyzed and built based on orthogonalizing against the columns on the left. This view also preserves the most recent information added to the system [6, 7]. One can still update the QR decomposition using Givens rotations as described in [65, Section 12.5.2] with this ordering and with columns sometimes removed from the various positions in the matrix, although it requires more operations. In the present context of numerical PDEs, however, we generally have one linear solve per iteration, and we have found even rebuilding a QR decomposition using efficient MATLAB or Python routines generally takes less that 10% (often around 1%) of the time for a linear solve.*

Additional results in [53] discuss choosing which set of secant equations to use and present methods that interpolate between Anderson Type-I/II and Broyden Type-I/II methods. A relation between Type-I AA and the Arnoldi method for linear problems is shown in [123].

2.2 ▪ Local convergence

The first convergence theory for AA can be found in [120], with refinements presented in [32, 77]. In contrast to the discussion above on the connection to quasi-Newton methods, the theory in this section is based on Algorithm 1.2. Here we recall three key results. The first is for linear problems, and the second two are for sufficiently smooth nonlinear contractive problems, first for depth $m = 1$ and then for general depths $m \geq 1$. Throughout the results, the correspondence between the fixed-point problem $g(x) = x$ and the zero-finding problem $f(x) = 0$ is $f(x) = g(x) - x$, and the simplifying assumption that the relaxation factor $\beta_k = \beta = 1$ is made throughout.

The first result is for AA applied to linear problems. Before stating the result, we recall the meanings of q-linear and r-linear convergence. If a sequence $\{s_k\}$ converges q-linearly to zero with q-factor c, it means there is a constant $0 < c < 1$ such that $\|s_{k+1}\| \leq c\|s_k\|$. If a sequence $\{s_k\}$ converges r-linearly to zero with convergence factor $0 < c < 1$, it means $\limsup_{k\to\infty} \|s_k\|^{1/k} \leq c$. In either case, if the bound holds with $c = 0$, the convergence is said to be superlinear.

Theorem 2.2 ([77, Theorem 5.2], [120, Theorem 2.1]). *Let M be an $n \times n$ matrix with $\|M\| = c < 1$. Let $m \geq 0$. Then AA(m), when applied to $g(x) = Mx + b$, converges to the solution $x^* = (I - M)^{-1}b$. Moreover, the residuals $f(x) = b - (I - M)x$ converge to zero q-linearly with a q-factor no greater than c.*

A motivating idea behind the transition from linear to nonlinear problems is similar to one used in the analysis of Newton's method: the linear terms in a Taylor expansion dominate close enough to the root. The local convergence theory for nonlinear problems assumes the following.

Assumption 2.3. *The function g has a fixed point x^*.*

- *g is continuously differentiable in the ball $B(x^*, \hat{\rho}) = \{x \mid \|x - x^*\| \leq \hat{\rho}\}$ for some $\hat{\rho} > 0$.*
- *There is some $c \in (0, 1)$ such that for all $x, y \in B(x^*, \hat{\rho})$, $\|g(x) - g(y)\| \leq c\|x - y\|$.*

In comparison to the assumptions made in Chapter 4, the first requirement is weaker and the second is stronger. Assumption 2.3 requires less regularity than the corresponding Assumption 4.1 (there is no Lipschitz requirement on g'). But, the corresponding Assumption 4.2 makes a more general assumption which is satisfied if g is contractive (required by Assumption 2.3) but does not require it.

The next result is for AA(1). Here it is assumed that the optimization is performed with the l^2 norm, and the closed form of the optimization is explicitly used in the proof. The result shows local q-linear convergence of the residuals.

Theorem 2.4 ([77, Theorem 5.4], [120, Theorem 2.4]). *Let Assumption 2.3 hold. If $\|x_0 - x^*\|$ is sufficiently small, the AA(1) residuals with l^2-optimization converge q-linearly. Moreover,*

$$\limsup_{k\to\infty} \frac{\|f(x_{k+1})\|}{\|f(x_k)\|} \le c, \tag{2.2.1}$$

where c is the Lipschitz constant from Assumption 2.3.

The last result in this section is local r-linear convergence of the residual, assuming the l^1 norm of the optimization coefficients remains bounded (either by assumption or by "arrangement"). This version of the theorem from [77] refers to a refinement of the theorem in [120] as developed in [32].

Remark 2.5. *We note a different convention in notation between the above literature and this book: we refer to the optimization coefficients in Algorithm 1.2 used in the update of x_{k+1} as α_j^{k+1}, where j runs from $k - m_k$ to k, whereas in the above literature they are α_j^k, where j runs from 0 to m_k. For consistency, we state the result here using our indexing.*

Theorem 2.6 ([32, Theorem 2.2], [77, Theorem 5.5], [120, Theorem 2.3]). *Let Assumption 2.3 hold. Assume that there is M_α such that, for all $k \ge 0$,*

$$\sum_{j=k-m_k}^{k} |\alpha_j^{k+1}| \le M_\alpha.$$

Then if the initial iterate x_0 is close enough to x^*, the AA(m) iterations converge, and

$$\limsup_{k\to\infty} \left(\frac{\|f(x_k)\|}{\|f(x_0)\|}\right)^{1/k} \le c. \tag{2.2.2}$$

These groundbreaking results are the first convergence results for AA, which had been in use more than 50 years from their first publication. They inspired many further investigations, including those of the current authors.

One particular idea that led to the results of Chapter 4 is the difference in hypotheses between Theorems 2.4 and 2.6, specifically that there is no bound required on the optimization coefficient for AA(1). This follows from using an explicit solution to the l^2 or least-squares optimization problem which is easily written out for $m = 1$. However, the solution of the optimization problem for $m > 1$ can also be written out explicitly, for instance using Gram-Schmidt or a QR factorization. It is following this line of thought that led to the analytical use of the entries in R of the QR factorization of the matrix in the least-squares problem (using Algorithm 1.1) in those results.

2.3 ▪ Connections to GMRES

For linear problems, AA is closely related to (but not the same as) the generalized minimal residual method (GMRES) [104]. Here we will highlight the most well known result relating the two: As shown in [123], AA is "essentially equivalent" to GMRES for linear problems, which means under certain assumptions "the iterates produced by either method can be readily obtained from those of the other method." One of these assumptions is that AA (and likewise GMRES) is run with "infinite memory" ($m_k = k$ for all k), which is not how the method is generally used in practice. As above, these results consider the case where the damping parameter $\beta_k = 1$ for all k (no relaxation).

For reference, we include the GMRES Algorithm from [104], as it appears there for the problem $n \times n$ problem $Ax = b$. Here $(\cdot, \cdot)$ denotes the Euclidean inner product.

ALGORITHM 2.1. GMRES(m).

Choose initial iterate x_0 and algorithmic depth parameter m
Compute $r_0 = b - Ax_0$ and $v_1 = r_0 / \|r_0\|$. ▷ $k = 0$
1: **for** $k = 1, \ldots$ **do** ▷ $k > 0$
2: Compute $h_{i,k} = (Av_k, v_i)$, $i = 1, 2, \ldots, k$
3: Set $\hat{v}_{k+1} = Av_k - \sum_{i=1}^{k} h_{i,k} v_i$
4: Set $h_{k+1,k} = \|\hat{v}_{k+1}\|$ and $v_{k+1} = \hat{v}_{k+1} / h_{k+1,k}$
5: Update: $x_k = x_0 + V_k y_k$, where y_k minimizes $J(y) = \|\beta e_1 - \bar{H}_k y\|$
6: **end for**

For the optimization problem in line 7 of Algorithm 2.1, $\bar{H}_k$ is H_k with an additional row appended whose only nonzero entry is $h_{k+1,k}$, and the vector e_1 is the standard Euclidean basis vector in $\mathbb{R}^{k+1}$, which coincides with the first column of the $(k+1) \times (k+1)$ identity matrix. In [104], the choice $\beta = \|r_0\|$ is made for convenience.

The assumptions under which the correspondence between the two algorithms is established are as follows.

Assumption 2.7.

- *$g(x) = Ax + b$ for $A \in \mathbb{R}^{n \times n}$ and $b \in \mathbb{R}^n$.*
- *AA (Algorithm 1.2) is not truncated, i.e, $m_k = k$ for each k.*
- *$(I - A)$ is nonsingular.*
- *GMRES is applied to $(I - A)x = b$ with the same initial point x_0 used to start AA.*

The following theorem compares AA($m_k = k$) iterates with iterates of GMRES applied to the linear system $(I - A)x = b$, starting with the same initial x_0. Iterates of AA (using Algorithm 1.2) are denoted by x_k^{AA}, and iterates of GMRES are denoted by x_k^{GMRES}. We point out the same change in the indexing of the optimization coefficients α_j^k as in Remark 2.5, so that the theorem below applies to the indexing convention used in Algorithm 1.2, which differs from the indexing used in [123].

Theorem 2.8 ([123, Theorem 2.2]). *Suppose that Assumption 2.7 holds and that, for some $k > 0$, $r_{k-1}^{GMRES} \neq 0$, and also $\|r_{j-1}^{GMRES}\| > \|r_j^{GMRES}\|$, for each j such that $0 < j < k$. Then*

$$\sum_{i=0}^{k} \alpha_{k-i}^{(k+1)} x_i^{AA} = x_k^{GMRES}, \quad \text{and } x_{k+1}^{AA} = g(x_k^{GMRES}).$$

A similar result is shown for the "essential equivalence" between Type-1 AA (2.1.16) and the Arnoldi, or full orthogonalization method (FOM) [103]. GMRES while originally posed for linear problems can be extended to nonlinear problems and is generally referred to then as nonlinear GMRES (NGMRES) [125]. We refer readers also to [113] for a useful discussion on this method and how it may be interpreted with respect to acceleration and preconditioning.

2.4 ▪ Some topics this book does not cover

The recent literature on AA has grown extensively in the last decade (as of this writing). We mention below several important topics that are beyond the scope of this book, and note that this is not an exhaustive list.

2.4.1 ▪ AA for linear problems

In addition to the results above in Section 2.3, specifics on AA for linear problems have been studied by a number of authors. The asympototic convergence rate of AA in comparison to related methods is discussed in [39, 124]. The classification of AA as a Krylov space method is studied in [94]. Additional closely related quasi-Newton methods may also be found in [70].

2.4.2 ▪ AA for nonsmooth problems

The convergence theory discussed here (see Chapter 4) assumes the fixed-point operator is sufficiently smooth that its Taylor expansion with a second-order integral remainder term can be written down. However, AA can also be used for problems with less smoothness. Convergence results for AA applied to a nonsmooth class of fixed-point problems are developed in [17, 18]. Another approach using a safeguarded version of Type-I AA (in the sense described in section 2.1) can be found in [128].

2.4.3 ▪ Connections to vector extrapolation methods

Vector extrapolation methods grew out of vector generalizations of the scalar-based Aitken's acceleration, also known as Aitken's delta-squared process which dates back to the 1920's [2]. Useful reviews of vector extrapolation methods can be found in the book [26] and the review paper [111]. The connection between two important classes known as *minimal polynomial* and *reduced rank* methods can be found in [110]. The literature on extrapolation methods and their analysis is extensive, especially compared with AA. Of particular interest is the connection between AA and vector extrapolation methods, analyzed via Shanks transformations, in [27]. Finally we will mention that a closer analogue to vector extrapolation methods is "restarted" AA, which differs from the "moving window" technique that we consider standard in this presentation.

2.4.4 ▪ Optimal parameter selection

The performance of AA depends on its implementation. In particular, how and how well it works on a given application strongly depends on how the parameters including the relaxation factors β_k and algorithmic depth m (or m_k) are chosen. Our approach here is to give strong heuristics of how to choose these parameters based on the theory rather than to give fixed prescriptions of how to set them.

It has been observed [91, 120] that AA reduces the sensitivity of the iteration to the relaxation factor. On one hand, the explanation is quite simple. For the fixed point iteration $x_{k+1} = g(x_k)$ written in terms of an update step $g(x_k) = x_k + w_{k+1}$, the update with relaxation is

$x_{k+1} = x_k + \beta_k w_{k+1}$, whereas the accelerated update (see line 8 of Algorithm 1.1 and (1.2.3)) is $x_k^\alpha + \beta_k w_{k+1}^\alpha$. In the base case

$$\frac{\partial}{\partial \beta_k} x_{k+1} = w_{k+1},$$

and in the accelerated case

$$\frac{\partial}{\partial \beta_k} x_{k+1} = w_{k+1}^\alpha.$$

From (1.2.7), we have $\left\|w_{k+1}^\alpha\right\| = \theta_{k+1} \left\|w_{k+1}\right\|$ with $\theta_{k+1} \leq 1$, and in general the inequality is strict. This reduction in sensitivity alone does not show how to pick β_k, however. One heuristic of choosing β_k based on θ_k is given in [50, Section 5.2] as $\beta_k = \beta_{max} - \theta_k/2$, where in that instance $\beta_{max} = 0.9$. This strategy adjusts the relaxation factor linearly with the gain factor to scale β_k between a chosen $\beta_{max} \leq 1$ and $\beta_{max} - 1/2$. A more modern approach is taken in [61], an in-depth study from an application in modeling magnetically confined fusion plasmas, which gives a thorough study of constant, heuristic, adaptive, and machine learning strategies for parameter selection. One of the interesting outcomes of [61] is that definite improvements in performance can be gained by choosing parameters well, but there may be various parameter combinations that result in "best" or near-best performance.

2.4.5 ▪ Preconditioning

As suggested in [29], AA could be better understood within a systematic and broad framework of left and right preconditioning for nonlinear solvers. This area has substantial room for development.

2.5 ▪ Software packages

AA can be implemented by practitioners by building it into existing codes as a post-processing procedure or by implementing it directly into newer codes. Additionally, a number of software packages include robust and efficient implementations of AA. These include the following:

- The KINSOL package within SUNDIALS [62] (https://computing.llnl.gov/projects/sundials), and thus other major software packages that link to this such as deal.II for HPC finite element computations [1].
- PETSc [12] (https://petsc.org)
- Julia package and accompanying notebooks (https://github.com/ctkelley) for the book Solving Nonlinear Equations with Iterative Methods: Solvers and Examples in Julia, C. T. Kelley [76].
- daarem (https://cran.r-project.org/web/packages/daarem/) an R package implementing damped Anderson acceleration with restarts for accelerating expectation-maximization (EM) and EM-Like Monotone Algorithms [72].
- Gaussian (https://gaussian.com/), a computational chemistry software package.

Additionally, MATLAB and Python codes are available with this book at https://epubs.siam.org/doi/book/10.1137/19781611978490#supplementary-material.

Chapter 3

Application problem descriptions

A little nonsense, now and then, is relished by the wisest men.
—Willy Wonka

We give several test examples in this book, as one of our goals in writing the book is to give practitioners a substantial range of examples which they can extend to their problems of interest. In this chapter we give details on all the PDEs (and one ODE) used throughout the book, including their discretizations, relevant solution properties, and the specific iterations we will use to solve them. Codes for each of these systems can be found in the code repository at https://epubs.siam.org/doi/book/10.1137/19781611978490#supplementary-material.

3.1 ▪ Discretization preliminaries

Several of our test PDEs will be discretized with the finite element method. For simplicity, we generally discuss implementation using conforming C^0 finite element implementations and using Dirichlet boundary conditions, since the solvers and in particular the application of AA are mostly independent of the specific discretization and boundary conditions. We assume domains are connected and have smooth or polygonal boundaries. Domains will be denoted by Ω.

We use the notation $(\cdot,\cdot)$ to denote the $L^2(\Omega)$ inner product $(u,v) := \int_\Omega uv\,dx$ and $\|\cdot\|$ to represent the $L^2(\Omega)$ norm $\|u\| = \sqrt{(u,u)}$. Other norms will be used in our analysis and will be distinguished using subscripts.

We define the natural function spaces and element spaces as follows. We begin with H_0^1, which is a common space for many PDE systems we consider:

$$X := H_0^1(\Omega) = \{v \in H^1(\Omega),\ v|_{\partial\Omega} = 0\}, \tag{3.1.1}$$

where

$$H^1(\Omega) := \{v \in L^2(\Omega), \nabla u \in L^2(\Omega)^d\},$$

and $L^2(\Omega)$ is the space of square integrable functions over the domain Ω. For the interested reader, a concise discussion of L^p and Sobolev spaces with their relevant properties can be found in [24, Chapter 1], in the context of fluids in [80, Chapter 1], or in a functional analytic setting in [51, Appendix D]. For the case of vector functions, we use the same notation $X = H_0^1(\Omega)^d$.

We suppose the domain is triangulated by a simplicial mesh (triangles in 2D or tetrahedra in 3D), and we assume the mesh, denoted by $\tau_h(\Omega)$, is regular and conforming, i.e., the diameters of the elements are close to uniform and the edges/faces of adjacent elements align; see, e.g.,

[24, 49] for more details on meshes. The notation $P_k(\tau_h)$ denotes the space of continuous, piecewise degree k polynomials on each element of τ_h. Often we will use the notation P_k and drop the τ_h when there is no confusion about which mesh is being used. The degree k will be specified for each test and ranges from 1 to 4 in our tests.

The finite element space $X_h \subset X$ is defined by

$$X_h := \{v \in P_k(\tau_h) \cap C^0(\overline{\Omega})\}. \tag{3.1.2}$$

The same notation is used for vector finite element functions. All of our tests use the same degree polynomials for the components of the vector functions.

Three of our test problems involve incompressible fluid flow: Navier-Stokes equations, the Boussinesq system for non-isothermal flow, and the Bingham equations for viscoplastic flow. These are all very widely studied systems, but more importantly for our purposes, their behaviors with nonlinear solvers are all quite different. For these problems, the natural velocity space is X given in (3.1.1), and the natural pressure space is

$$Q := L_0^2(\Omega) = \{q \in L^2(\Omega), \int_\Omega q\, dx = 0\}.$$

The divergence-free subspace of X is defined by

$$V := \{v \in X, \ (\nabla \cdot v, q) = 0 \ \forall q \in Q\}.$$

The pair of spaces X, Q satisfies the inf-sup condition [64], one of the essential conditions for establishing well-posedness of problems of this form: there exists a constant $C > 0$ satisfying

$$0 < C < \inf_{q \in Q} \sup_{v \in X} \frac{(\nabla \cdot v, q)}{\|\nabla v\| \|q\|}. \tag{3.1.3}$$

In contrast to the related coercivity estimate used for equations written in second-order form, the discrete analogue of (3.1.3), known as the discrete inf-sup condition, does not automatically hold for arbitrary subspaces of X and Q. Discrete subspace pairs $X_h \subset X$ and $Q_h \subset Q$ that do satisfy the discrete inf-sup condition are known as inf-sup stable pairs. Further discussion on the inf-sup condition and the discrete inf-sup condition can be found in [24, Chapter 12], and [80, Chapter 4] in the context of fluid flow; see also the Babuska-Brezzi condition.

Finite element pressure spaces Q_h will take one of two forms. For continuous pressures (in which case $Q_h \subset H^1(\Omega)$),

$$Q_h := \{v \in P_j(\tau_h) \cap C^0(\overline{\Omega}\}.$$

For this case, Q_h is often referred to in the literature (and in this book) by P_j. Discontinuous pressures will also be used, and they are defined by

$$Q_h := \{v \in P_j(\tau_h) \cap Q\}.$$

Here, Q_h is often referred to as P_j^{disc}.

We will use conforming and inf-sup stable velocity-pressure pairs in our tests, and in particular we use the (P_2, P_1) Taylor-Hood element and the (P_k, P_{k-1}^{disc}) Scott-Vogelius element with $k = d$ (see, e.g., [24, 69, 75] for more on these element choices). Taylor-Hood elements are known to be inf-sup stable on any triangulation or tetrahedralization [24, 80], with the C from (3.1.3) potentially depending on mesh properties such as a minimum angle condition. Scott-Vogelius elements, on the other hand, may not be inf-sup stable on general triangulations or tetrahedralizations, but when $k = d$ they are when the mesh is constructed as a barycenter refinement (Alfeld split in the Guzman/Neilan vernacular) of a triangulation or tetrahedralization [9, 69, 75, 129].

The discretely divergence-free subspace is defined by

$$V_h := \{v \in X_h,\ (\nabla \cdot v, q) = 0\ \forall q \in Q_h\}.$$

For Scott-Vogelius elements, the space V_h is strongly divergence-free in the sense that $\|\nabla \cdot v_h\| = 0$ for all $v_h \in V_h$.

3.2 ▪ Nonlinear Helmholtz equation

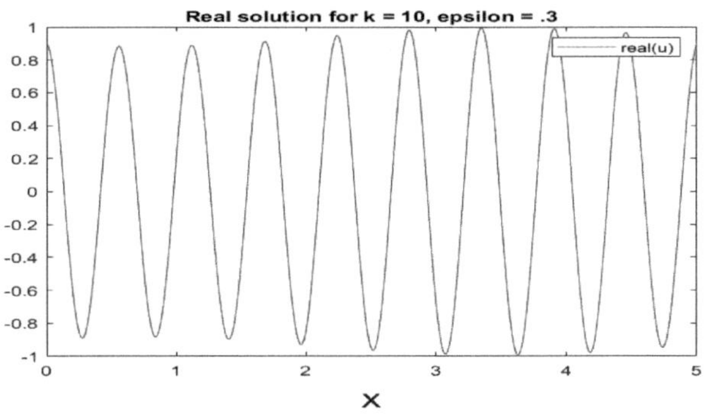

Figure 3.1. *The plot above shows the real part of the solution of the nonlinear Helmholtz problem with $k = 10$ and $\epsilon = 0.3$ on the left half of the domain.*

The first test problem we will discuss is the nonlinear Helmholtz (NLH) equation from optics, where the interest is in the propagation of continuous-wave laser beams through transparent dielectrics. The NLH system is written in 1D on interval $[0, 10]$ as follows: Find $u : [0, 10] \to \mathbb{C}$ satisfying

$$u_{xx} + k^2 \left(1 + \epsilon(x)|u|^2\right) u = 0, \quad 0 < x < 10,$$

where $u = u(x)$ represents an unknown complex scalar electric field, k is the linear wavenumber, and $\epsilon(x)$ is a material dependent quantity depending on the linear index of refraction and Kerr coefficient. The physically consistent two-way boundary condition is derived in [15, 56] and is given by

$$\begin{aligned} u_x + iku &= 2ik \text{ at } x = 0, \\ u_x - iku &= 0 \text{ at } x = 10. \end{aligned}$$

Despite this nonlinear boundary value problem being 1D (and therefore an ODE), this system can be quite challenging for nonlinear solvers due to its cubic nonlinearity [14, 15]. We include this "simpler" system because it provides an example for fast testing of codes and new ideas. A plot of the real part of a NLH solution with $k = 10$ and $\epsilon = 0.3$ is shown in Figure 3.1.

A Picard iteration for the NLH system takes the form

$$(u_{j+1})_{xx} + k^2 u_{j+1} + k^2 \epsilon(x)|u_j|^2 u_{j+1} = 0, \qquad 0 < x < 10, \tag{3.2.1}$$

$$(u_{j+1})_x + iku_{j+1} = 2ik, \quad x = 0, \tag{3.2.2}$$

$$(u_{j+1})_x - iku_{j+1} = 0, \qquad x = 10. \tag{3.2.3}$$

This iteration (3.2.1)-(3.2.3) is attractive because of its simplicity and because it works directly with complex numbers. The Newton iteration, on the other hand, requires decomposition into the real and imaginary parts and thus leads to a fully coupled block linear system at each iteration.

We discretize in space using $N = 2000$ equally spaced points $\{0 = x_1, x_2, \ldots, x_N = 10\}$ and a second order finite difference approximation for $n = 1 : N-1$:

$$(u_{j+1})_{xx}(x_n) \approx \frac{u_{j+1}(x_{n+1}) - 2u_{j+1}(x_n) + u_{j+1}(x_{n-1})}{h^2},$$

where h is the spacing of the discretization points. Second-order approximation is also used for the boundary conditions with the usual 3 point stencil. For more on finite difference stencils, see, e.g., [10].

The fixed-point operator g for this problem is thus the solution operator to the finite difference approximation to the Picard iteration (3.2.1)-(3.2.3), meaning that given u_j, we can express solving (3.2.1)-(3.2.3) for u_{j+1} as $u_{j+1} = g(u_j)$. For an initial guess, we use the nodal interpolant of

$$u_0 = (\cos(kx) + i\sin(kx)),$$

i.e., the linear ($\epsilon = 0$) Helmholtz equation solution for the same k.

For our tests, we consider various parameters k and ϵ and note that with larger values of each the NLH problem is harder due to larger emphasis of the nonlinear term. AA was used to improve convergence to this iteration for NLH in [89]. NLH is used as an example problem in Chapters 4 and 5. Codes for AA applied to this system are given in the code repository for both Algorithm 1.1 (NLH_AA2.m) and Algorithm 1.2 (NLH_AA.m).

3.3 ▪ Navier-Stokes equations

The NSE make a great test problem for solvers, since they are physically relevant, widely used, and notoriously hard to solve. We will consider both 2D and 3D test problems. The systems we consider take the form of

$$\begin{aligned} u\cdot\nabla u + \nabla p - Re^{-1}\Delta u &= f, \\ \nabla\cdot u &= 0, \\ u|_{\partial\Omega} &= u_{bc}, \end{aligned}$$

where u and p are the unknown velocity and pressure, f is a given external forcing, Re is the Reynolds number, and u_{bc} is a given (Dirichlet) boundary condition.

The two NSE test problems we consider are the 2D driven cavity and 3D driven cavity, which are both very commonly used for testing solvers, including for AA tests in [50, 91, 88]. They both use no forcing ($f = 0$) and no slip velocity everywhere except the top, and on the top the lid moves with constant velocity 1 in the x-direction. Solution plots are shown in Figure 3.2. Most solvers work well for lower Re, but convergence gets slower as Re increases. For Re large enough, most solvers fail. Our tests will show how AA will allow solvers to converge for higher Re than would otherwise be possible.

The Picard iteration (sometimes called the Oseen linearization) for the steady NSE is given in weak form by the following: Given $u_j \in V$, find $(u_{j+1}, p_{j+1}) \in (X, Q)$ satisfying for all $(v, q) \in (X, Q)$,

$$\begin{aligned} (u_j\cdot\nabla u_{j+1}, v) - (p_{j+1}, \nabla\cdot v) + Re^{-1}(\nabla u_{j+1}, \nabla v) &= (f, v), \\ (\nabla\cdot u_{j+1}, q) &= 0. \end{aligned}$$

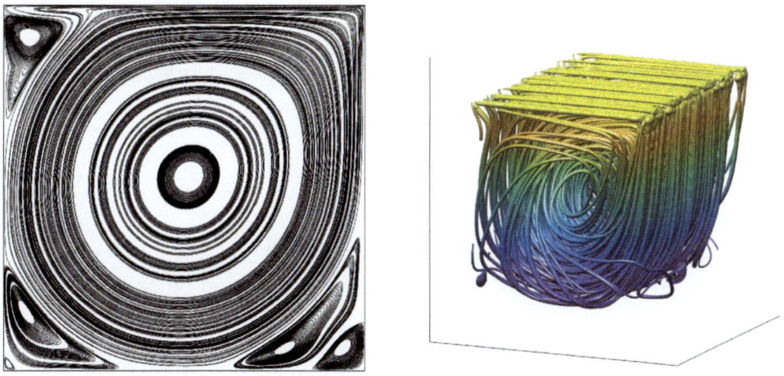

Figure 3.2. *The plot above shows the solution of the steady Navier-Stokes for the 2D driven cavity problem at Re =10,000 (left) and 3D driven cavity problem at Re =1600.*

In our tests we use an initial guess of $u_0 = 0$, which implies that u_1 is the Stokes solution with the given problem data. The solution operator of this problem, written as $u_{j+1} = g(u_j)$, is proven to be well-posed in [91].

We will use (P_2, P_1) Taylor-Hood elements or (P_k, P_{k-1}^{disc}) Scott-Vogelius elements (with $k = d$ and barycenter refined meshes) for our tests. In both cases, the Picard scheme above can be repeated with X, Q, V replaced by X_h, Q_h, V_h with one exception: For the convection term, when Taylor-Hood elements are used, since $u_h \in V_h$ does not yield $\nabla \cdot u_h = 0$ we employ skew-symmetry via [80, 119]

$$u_j \cdot \nabla u_{j+1} \to u_j \cdot \nabla u_{j+1} + \frac{1}{2}(\nabla \cdot u_j)u_{j+1}.$$

This modification, originally developed by Temam [117, 118], is mathematically consistent since it arises from adding $0 = -\frac{1}{2}(\nabla \cdot u)u$ to the PDE. This is important to the analysis as well as computations.

The test problems we use are the lid-driven cavity benchmark problems in both 2D and 3D. In 2D, the domain is the unit square, there is no forcing ($f = 0$), and the Dirichlet boundary conditions are set to be 0 on all sides but $\langle 1, 0\rangle^T$ at the top, to represent the moving lid. The 3D test problem is analogous, with the boundary at the top enforced to be $\langle 1, 0, 0\rangle^T$. Tests will be run with various meshes, AA depths, AA relaxation parameters, and Reynolds numbers. AA is shown in [50, 88, 91] to have a significant positive impact on solvers for the NSE. The NSE are used as an example problem in Chapters 4 and 5.

There are separate codes in the repository for 2D and 3D Navier-Stokes. For the 3D problem we use Algorithm 1.1, the driver file is the following: NSE_AndersonPicard_3DCavity.m, and the driver NSE_AndersonPicard_3DCavity_Filtering.m also allows for both filtering and PCA. For the 2D problem, we use Algorithm 1.2 and the driver file is the following: NSE_Anderson Picard_2DCavity.m. These codes allow for easy changing of AA parameters, Reynolds numbers, and meshes, and for the filtering code the filtering parameters as well.

3.4 ▪ Gross-Pitaevskii equations

Another interesting test problem is the Gross-Pitaevskii equations (GPE). The GPE system is a model for stationary solutions of the nonlinear Schrödinger (NLS) equation, which is also commonly referred to as the non-rotational GPE in the context of Bose-Einstein condensates

(BEC) [13, 79, 86]. The GPE are given by

$$\mu\phi(x) = -\frac{1}{2}\Delta\phi(x) + v(x)\phi(x) + \eta|\phi(x)|^2\phi(x), \quad x \in \Omega,$$
$$\phi(x) = 0, \quad x \in \partial\Omega,$$
$$\int_\Omega |\phi(x)|^2 \, dx = 1.$$

Here, v is a given trapping potential of the form $v(x) = \frac{1}{2}(\gamma_1^2 x_1^2 + \cdots + \gamma_d^2 x_d^2)$ with $\gamma_i > 0 \; \forall i$ and real parameter η, ϕ is the unknown, and the eigenvalue μ can be calculated as

$$\mu = \int_\Omega \left(\frac{1}{2}|\nabla\phi|^2 + v|\phi|^2 + \eta|\phi|^4 \right) dx.$$

In the GPE system, ϕ represents the macroscopic wave function of the condensate. The sign of parameter η (positive/negative) represents attraction/repulsion of the condensate atoms in GPE.

For a solver, we use the Picard-projection fixed-point iteration from [57], denoted by PP$_h$, which is defined by the following:

PP$_h$ Step 1: Given $\phi_k \in X_h$ with $||\phi_k|| = 1$, find $\hat{\phi}_{k+1} \in X_h$ satisfying

$$\frac{1}{2}(\nabla\hat{\phi}_{k+1}, \nabla\chi) + (v\hat{\phi}_{k+1}, \chi) + \eta(|\phi_k|^2\hat{\phi}_{k+1}, \chi) = (\phi_k, \chi).$$

PP$_h$ Step 2: Calculate

$$\phi_{k+1} = \frac{\hat{\phi}_{k+1}}{||\hat{\phi}_{k+1}||}.$$

It is the combination of Steps 1 and 2 that defines the fixed-point operator for this problem:

$$\phi_{k+1} = g(\phi_k).$$

It is straightforward to prove that g is well-defined [57].

We will solve GPE for a 2D test from [13] that represents a harmonic oscillator potential together with a potential from a stirrer corresponding to a far-blue detuned Gaussian laser beam:

$$v(x, y) = \frac{1}{2}\left(x^2 + y^2\right) + 4e^{-\left((x-1)^2 + y^2\right)}.$$

Solutions are computed on $\Omega = (-8, 8)^2$ using varying η, with initial guess

$$\phi_0(x, y) = \frac{1}{\pi^{1/2}} e^{-(x^2+y^2)/2},$$

using a uniform 128x128 triangulation and globally continuous P_2 elements. A plot of the solution with η =10 and 11,000 is shown in Figure 3.3 as a surface plot. Codes for the GPE problem are in the repository, and the driver file is the following: GPEPicard_Filtering.m. Length and angle filtering can be used here, as well as TSVD filtering. Algorithm 1.1 is used for the Anderson acceleration. The GPE are used as an example problem in Chapter 5.

3.5 ▪ A monotone quasilinear equation

Our next example is a finite element discretization of a monotone quasilinear problem of the form

$$-\operatorname{div}(\mu(|\nabla u|)\nabla u) = f \text{ in } \Omega = (0,1) \times (0,1), \quad u = 0 \text{ on } \partial\Omega. \tag{3.5.1}$$

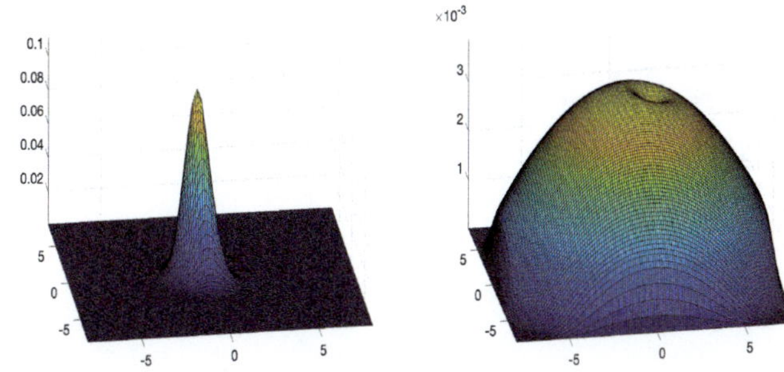

Figure 3.3. *Shown above is a surface plot of the converged ground state Gross-Pitaevskii solution using* η *=10 (left) and* η *=11,000 (right). [57], distributed under CC BY 4.0 DEED License (https://creativecommons.org/licenses/by/4.0/).*

Problems of this form arise in stationary conservation laws including electrostatics, magnetostatics, elasticity, and heat conduction [60]. Here we consider a particular instance of (3.5.1) describing steady nonlinear diffusion, adapted from [36, Example 1], and further discussed in [89], with

$$\mu(|\nabla u|) = 1 + \arctan(|\nabla u|), \quad f = \pi. \tag{3.5.2}$$

A numerical solution to (3.5.1)-(3.5.2) is shown in Figure 3.4 on the left. As shown in [36], and see also [30] and [127], the solution u to (3.5.1)-(3.5.2) agrees with the solution to the following weak form: Find $u \in X$ given by (3.1.1) such that

$$(\mu(|\nabla u|)\nabla u, \nabla v) = (f, v) \text{ for all } v \in X. \tag{3.5.3}$$

The discrete form of (3.5.3) is the following: Find $u \in X_h$ given by (3.1.2) such that

$$(\mu(|\nabla u|)\nabla u, \nabla v) = (f, v) \text{ for all } v \in X_h. \tag{3.5.4}$$

We consider AA applied to a Picard iteration for this problem as a means of reducing the iteration count and relaxing (or removing!) the dependence of convergence on damping. Starting with an initial iterate u_0, a Picard iteration for (3.5.3) can be defined in terms of an update step w_{k+1}, found by the solution to the following linear problem: Find $w_{k+1} \in X_h$ such that

$$(\nabla w_{k+1}, \nabla v) = (f, v) - (\mu(|\nabla u_k|)\nabla u_k, \nabla v) \text{ for all } v \in X_h. \tag{3.5.5}$$

By the analysis of [36], the fixed-point iteration $g(u_k) = u_k + \beta w_{k+1}$ is contractive for $0 < \beta \leq \beta^* := (1 + \sqrt{3}/2 + \pi/3)^{-2}$. We will consider iterations both with and without the damping factor β sufficiently small.

Figure 3.4. *Left: solution to the monotone quasilinear problem* (3.5.1)-(3.5.2). *Right: Regularized p-Laplace solution with* $p = 1.06$, $f = 2$, *and* $\epsilon = 10^{-14}$.

This example is used in Chapter 5. Codes for this problem are in the repository, and the driver file is the following: strmono.py.

3.6 ▪ p-Laplace

The elliptic p-Laplace (or p-Poisson) equation

$$-\operatorname{div}\left((|\nabla u|^2/2)^{(p-2)/2}\nabla u\right) = f \text{ in } \Omega$$

arises as a model problem in non-Newtonian flow, turbulent flow, and flow in porous media [46, 47], as well as in the discretization of abstract linear operators in Banach spaces [82]. It makes for another good test example for AA, as developed in [88, 89]. If $1 < p < 2$, the equation is singular, as the exponent on the nonlinear diffusion coefficient is negative and blows up as $|\nabla u|$ goes to zero. Hence we consider a regularized equation

$$-\operatorname{div}\left(\left(\epsilon^2 + \frac{1}{2}|\nabla u|^2\right)^{(p-2)/2}\nabla u\right) = f \text{ in } \Omega \text{ with } \epsilon = 10^{-14}. \tag{3.6.1}$$

Solving the equation for $1 < p < 2$ is more challenging when p is closer to 1. Similar to [29, 88], we consider (3.6.1) over the domain $\Omega = (0,2) \times (0,2)$ with homogeneous Dirichlet boundary conditions, constant forcing, and a relatively bad initial guess $u_0 = xy(x-1)(y-1)(x-2)(y-2)$. A numerical solution to (3.6.1) with $\epsilon = 10^{-14}$ and $f = 2$ is shown in Figure 3.4 on the right.

We consider applying AA to the Picard iteration for this problem both to enable convergence in the preasymptotic regime and to accelerate convergence in an asymptotic regime. The (discretized) Picard iteration is given by the following update step: Find $w_{k+1} \in X_h$ satisfying for all $v \in X_h$,

$$(a_{\epsilon,p}(u_k)\nabla w_{k+1}, \nabla v) = (f, v) - (a_{\epsilon,p}(u_k)\nabla u_k, \nabla v), \tag{3.6.2}$$

with $a_{\epsilon,p}(u_k) = \left(\epsilon^2 + \frac{1}{2}|\nabla u_k|^2\right)^{(p-2)/2}$. Hence the fixed-point iteration takes the form

$$u_{k+1} = g(u_k) = u_k + w_{k+1}.$$

We use this example in Chapters 4 and 5. The codes for this problem are in the repository, and the driver file is rpLap.py.

3.7 ▪ Bingham equations

A Bingham plastic is a material that is solid at lower shear stresses, but flows with a constant viscosity when larger shear stress is applied [19]. Examples include fresh concrete, dough, blood in the capillaries, toothpaste, muds, and ketchup [40] (in the case of ketchup, the force should be applied upward and at angle so the "57" on the bottle is hit directly).

The governing equations of Bingham plastics is

$$\begin{aligned} -\nabla \cdot \tau + \nabla p &= f, \\ \nabla \cdot u &= 0, \end{aligned} \tag{3.7.1}$$

in a bounded and connected domain $\Omega \subset \mathbb{R}^d$, $d = 2, 3$. Here, τ is the stress, u is the fluid velocity, and p is the pressure.

The behavior change of Bingham plastics occurs after the applied stress τ exceeds a certain threshold called the yield stress τ_s. Denote by $Du = \frac{1}{2}[\nabla Du + \nabla Du^T]$ the strain rate tensor defined as a symmetric part of velocity gradient, and define $|Du| = \sqrt{Du : Du}$ to be the Frobenius norm of Du. The applied stress τ in (3.7.1) is defined by

$$\tau = 2\mu Du + \tau_s \frac{Du}{|Du|}.$$

If $|\tau| > \tau_s$, a Bingham plastic behaves as a Stokes fluid, but if $|\tau| \leq \tau_s$, then Bingham plastic behaves as solid material.

To avoid the issue of having a fluid region and a solid region, and an interface between them, the Bercovier-Engelman regularization was developed to approximate Bingham solutions on a single domain. Define $|Du|_\varepsilon = \sqrt{Du : Du + \varepsilon^2}$ in the shear-dependent viscosity (which we call $\hat{\mu}$), with ε denoting the regularization parameter. The regularized system takes the form

$$-\nabla \cdot \left(2\mu + \frac{\tau_s}{|Du|_\varepsilon}\right) Du + \nabla p = f, \qquad \nabla \cdot u = 0. \tag{3.7.2}$$

Now, the entire domain is treated computationally as a single region, and the plug region can be recovered by inspecting regions of high viscosity. The obvious drawback is the physical inconsistency, and it is known from [68] that comparing the regularized solution to the nonregularized solution yields only

$$\|D(u - u_{nonreg})\| \leq C\sqrt{\epsilon}.$$

Thus not surprisingly, one must choose small ϵ for good accuracy [41, 58, 105], but unfortunately, small ϵ causes solvers to fail. A numerical solution to (3.7.2) with $\mu = 1$ and $\epsilon = 10^{-6}$ is shown in Figure 3.5.

We define the Picard iteration for regularized Bingham as follows. Given $u^0 \in V_h$, for $k = 1, 2, \ldots$, find $u_k \in V_h$ such that

$$2\mu(Du_k, Dv) + \tau_s \left(\frac{Du_k}{|Du_{k-1}|_\varepsilon}, Dv\right) = (f, v) \quad \forall v \in V_h. \tag{3.7.3}$$

The analysis of the Picard iteration (3.7.2) given in [8] demonstrates convergence but not global linear convergence. Numerical tests suggest it may be sublinearly convergent or perhaps linearly convergent far into the asymptotic range. Our tests herein will use Taylor-Hood and Scott-Vogelius finite elements. We will look at two test problems. The first is the 2D driven cavity

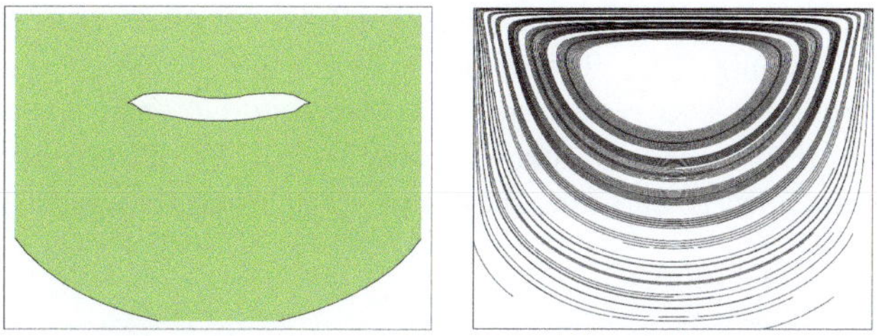

Figure 3.5. *Shown above are the stress (left, green is stress above yield stress $\tau_s = 1$) and velocity streamlines (right) for the solution to the driven cavity problem for regularized Bingham using $\mu = 1$ and $\epsilon = 10^{-6}$. [90], distributed under CC BY 4.0 DEED License (https://creativecommons.org/licenses/by/4.0/).*

problem, which uses the same setup as above for NSE. The second is 2D flow between two parallel plates, whose setup differs from the driven cavity only in its given Dirichlet boundary conditions. Codes for regularized Bingham can be found in the repository, and the driver file is BinghamRegDriver.m. The regularized Bingham tests appear in Chapters 4 and 8.

3.8 ▪ Boussinesq equations

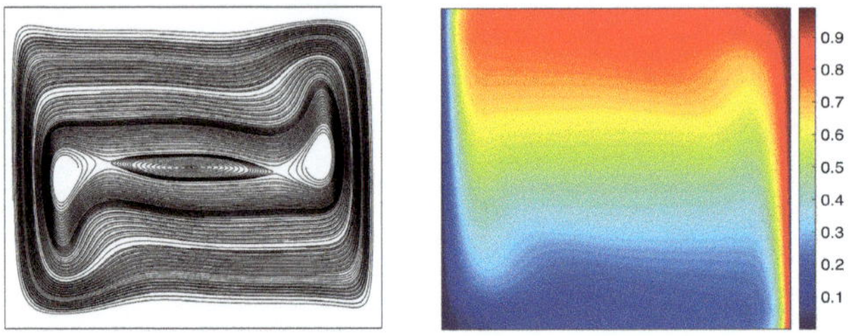

Figure 3.6. *The $Ra = 10^6$ Boussinesq velocity (left) and temperature (right) solutions are shown above for the differentially heated cavity problem.*

The Boussinesq equations describe non-isothermal flow, which typically are flows driven by buoyancy in applications such as ventilation, solar collectors, and window insulation, to name a few. The steady Boussinesq system takes the following form in a domain $\Omega \subset \mathbb{R}^d$ ($d = 2$ or 3):

$$\begin{aligned}(u \cdot \nabla)u - \nu\Delta u + \nabla p &= Ri(0;\theta) + f,\\ \nabla \cdot u &= 0,\\ (u \cdot \nabla)\theta - \kappa\Delta\theta &= \gamma,\end{aligned} \tag{3.8.1}$$

with u representing the velocity, p the pressure, θ the temperature (or density), and f and γ the external momentum forcing and thermal sources. The kinematic viscosity $\nu > 0$ is defined as the inverse of the Reynolds number ($Re = \nu^{-1}$), and the thermal conductivity $\kappa = Re^{-1}Pr^{-1}$, where Pr is the Prandtl number and Ri is the Richardson number. Appropriate boundary conditions are required to determine the system. The Rayleigh number is defined by $Ra = Ri \cdot Re^2 \cdot Pr$, and higher Ra leads to more complex physics as well as more difficulties in numerically solving the system. Velocity and temperature solution plots for (3.8.1) with $Ra = 10^6$ are shown in Figure 3.6.

For the Boussinesq system, we consider AA applied to both the Picard and the Newton iterations, as these are both commonly used to solve the system. The Newton iteration is locally quadratically convergent and thus is appropriate for lower Ra or when using a continuation method to solve for very large Ra. The Picard iteration is more efficient at each iteration, as it decouples the momentum-mass system from the temperature (density) equation, and effective preconditioners exist for these linear system [16, 48, 71, 20]. The Picard iteration is also stable, but will converge only linearly and thus may take many more iterations than Newton (when Newton converges).

The Newton iteration for this system takes the form

$$(u_{k-1} \cdot \nabla)u_k + (u_k \cdot \nabla)u_{k-1} - (u_{k-1} \cdot \nabla)u_{k-1} - \nu\Delta u_k + \nabla p_k = Ri(0;\theta_k) + f, \tag{3.8.2}$$

$$\nabla \cdot u_k = 0, \tag{3.8.3}$$

$$(u_{k-1} \cdot \nabla)\theta_k + (u_k \cdot \nabla)\theta_{k-1} - (u_{k-1} \cdot \nabla)\theta_{k-1} - \kappa\Delta\theta_k = \gamma, \tag{3.8.4}$$

together with appropriate boundary conditions. However, solving this system for large problems can be very difficult because it yields fully coupled linear systems. As the Rayleigh number increases, the steady solution undergoes bifurcations at certain values and in turn causes the Jacobian to become singular. For Ra near these points, we will discuss how AA can help Newton, even though AA will generally hurt quadratic convergence.

The Picard iteration for Boussinesq is given by

$$u_{k-1} \cdot \nabla u_k - \nu \Delta u_k + \nabla p_k = Ri(0; \theta_k) + f, \tag{3.8.5}$$

$$\nabla \cdot u_k = 0, \tag{3.8.6}$$

$$u_{k-1} \cdot \nabla \theta_k - \kappa \Delta \theta_k = \gamma. \tag{3.8.7}$$

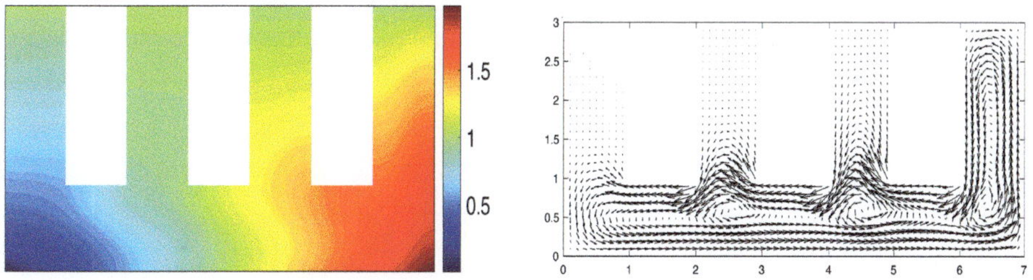

Figure 3.7. *Plots above show the temperature (left) and velocity (right) of the Boussinesq simulation with* $Ra = 2100$.

We consider two test problems for the Boussinesq equations. The first is the classical differentially heated cavity problem, for which the domain is the unit square, and no slip velocity is enforced on all walls. Perfect insulation is enforced $\nabla T \cdot n = 0$ on the top and bottom, and $T(1, y) = 1$, $T(0, y) = 0$ on the sides. The initial iterates $u_0 = 0$, $T_0 = 0$ are used for all tests. Tests are performed for varying Ra.

The second test problem is from [92], which is similar to the differentially heated cavity but on a more complex domain; see Figure 3.7. Here, we take no slip velocity on all walls but $T(x, 1) = 1$, $T(x, 0) = \frac{2x}{7}$, and $\nabla T \cdot n = 0$ on all other boundaries.

To display the weak form, we first define the temperature space by

$$W = \{\chi \in H^1(\Omega),\ \chi|_{\Gamma_1} = 0\},$$

noting that $\partial\Omega = \Gamma_1 \cup \Gamma_2$, where Γ_1 is the Dirichlet part of the boundary for temperature and the rest of the boundary, Γ_2, has perfect insulation, i.e., the homogeneous Neumann conditions. We assume $f, \gamma \in H^{-1}(\Omega)$. Then the weak form of the Picard scheme, with the forcing f and γ accounting for the inhomogeneous Dirichlet boundary conditions, reads as follows:

Step 1: Find $\theta_k \in W$ satisfying

$$(u_{k-1} \cdot \nabla \theta_k, \chi) + \kappa(\nabla \theta_k, \nabla \chi) = \langle \gamma, \chi \rangle \ \forall \chi \in W.$$

Step 2: Find $(u_k, p_k) \in (X, Q)$ satisfying for all $(v, q) \in (X, Q)$,

$$(u_{k-1} \cdot \nabla u_k, v) - (p_k, \nabla \cdot v) + \nu(\nabla u_k, \nabla v) = \langle f, v \rangle + Ri((0; \theta_k), v),$$
$$(\nabla \cdot u_k, q) = 0,$$

where $\langle \cdot, \cdot \rangle$ represents the $H^1 - H^{-1}$ duality pairing.

The Newton iteration's weak form is defined analogously as a single step: Find $(u_k, p_k, \theta_k) \in (X, Q, W)$ satisfying for all $(v, q, \chi) \in (X, Q, W)$,

$$
\begin{aligned}
(u_{k-1} \cdot \nabla u_k, v) + (u_k \cdot \nabla u_{k-1}, v) - (p_k, \nabla \cdot v) + \nu(\nabla u_k, \nabla v)& \\
= \langle f, v\rangle + Ri((0;\theta_k), v) + (u_{k-1} \cdot \nabla u_{k-1}, v)&, \\
(\nabla \cdot u_k, q) = 0&, \\
(u_{k-1} \cdot \nabla \theta_k, \chi) + (u_k \cdot \nabla \theta_{k-1}, \chi) + \kappa(\nabla \theta_k, \nabla \chi) = \langle \gamma, \chi\rangle + (u_{k-1} \cdot \nabla \theta_{k-1}, \chi)&.
\end{aligned}
$$

We discretize with finite elements, using Lagrange degree 2 elements $W_h (= P_2 \cap W)$ for temperature, and either Taylor-Hood or Scott-Vogelius elements for velocity and pressure. The finite element schemes are the same as the weak forms above but with W, X, Q replaced by W_h, X_h, Q_h. As above, when Taylor-Hood is used we employ skew-symmetry for the convective terms. Codes for AA applied to Picard and Newton can be found in the code repository, and the driver file is BoussDriverAA.m.

Chapter 4

One-step acceleration analysis

Clothes make the man. Naked people have little or no influence in society.
—Mark Twain

This chapter provides a one-step analysis of AA that shows how the lowest-order term in the residual bound is reduced, effectively improving the linear convergence rate of the iteration at the cost of introducing higher-order terms into the residual expansion. It follows closely the analysis from [88] of the authors, which sharpens our results from [50] and generalizes beyond contractive operators. The main theory is for a single step, and the improvement from AA at a single step is characterized by the improvement (gain) offered by the AA optimization problem at that step. The gain factor, defined by (1.2.7), ranges between 0 and 1 and scales the linear convergence rate of the fixed-point operator, but at the price of introducing higher-order terms. Our numerical tests reveal that this gain can change dramatically from step to step (0.9 on one step and 0.5 the next!). Because of this, we do not attempt to compute an asymptotic gain factor or accelerated convergence rate.

A fundamental aspect of this theory is the relation between the optimization coefficients and the gain from the optimization problem. Below, this relationship is understood through a QR factorization. Using Algorithm 1.1, the QR is a natural way to solve the least-squares problem computationally. We find it is a natural way to understand the algorithm theoretically as well.

4.1 ▪ Preliminaries and main assumptions

For the theory that follows in this chapter, we consider seeking a fixed point of Fréchet differentiable operator $g : X \to X$ for Hilbert space $X \subset \mathbb{R}^n$ equipped with inner product $(\cdot, \cdot)$ and induced norm $\|\cdot\|$.

We consider the AA algorithm with depth m applied to the fixed-point problem $g(x) = x$, where the stage-k differences between iterates e_k and residuals w_k are given by (1.2.1). The main analysis of this chapter will use the unconstrained optimization problem form of AA, i.e., Algorithm 1.1, where the least-squares optimization problem is given by

$$\gamma^{k+1} = \operatorname{argmin} \|F_k\gamma - w_{k+1}\| \,. \tag{4.1.1}$$

Introducing the notation

$$\delta w_{k+1} = w_{k+1} - w_k, \tag{4.1.2}$$

we recall from Algorithm 1.1 that the matrices E_k and F_k used in the optimization and update are given by

$$\begin{aligned} E_k &:= \begin{pmatrix} e_k & e_{k-1} & \cdots & e_{k-m_k+1} \end{pmatrix}, \\ F_k &:= \begin{pmatrix} \delta w_{k+1} & \delta w_k & \ldots & \delta w_{k-m_k+2} \end{pmatrix}. \end{aligned} \tag{4.1.3}$$

A thin (economy) QR decomposition of F_k will be explicitly used in the analysis to extract relations between the optimization gain θ_k and optimization coefficients γ^{k+1}. The vector of optimization coefficients γ^{k+1} of (4.1.1) can be expressed as the solution to $R\gamma^{k+1} = Q^T w_{k+1}$, where $F_k = QR$ is the thin QR decomposition of F_k, as in [65, Chapter 5].

A key repercussion of this approach is that assumptions on the boundedness of the optimization coefficients as required in [50, 91], and [77, 120] for $m > 1$, are replaced by assumptions on the sufficient linear independence between columns of F_k, and in particular on the angles between the columns of F_k. As shown in this and the next chapter, satisfaction of those assumptions can be enforced through a simple filtering method to remove these angular (near) linear dependencies in F_k. In the next chapter, we will show how to use this as part of a two-part filtering strategy to enforce good conditioning of the least-squares problem.

4.1.1 ▪ Smoothness assumption

Our first assumption is on the smoothness of fixed-point operator g.

Assumption 4.1. *Assume $g \in C^1(X)$ has a fixed point x^* in X, and there are positive constants κ_g and $\hat{\kappa}_g$ with the following:*

1. $\|g'(x)z\| \le \kappa_g \|z\|$ *for all* $x, z \in X$.
2. $\|g'(x)z - g'(y)z\| \le \hat{\kappa}_g \|x - y\| \, \|z\|$ *for all* $x, y, z \in X$.

A particular case of interest is finding a zero of a function $f : X \to X$, where the system of nonlinear equations $f(x) = 0$ comes from the discretization of a nonlinear PDE. Then $g(x) = f(x) + x$ converts between the zero-finding and fixed-point problems. If g satisfies Assumption 4.1, then f has a zero $x^* \in X$, $f \in C^1(X)$, and

$$\|f'(x)z - f'(y)z\| = \|(g'(x) - I)z - (g'(y) - I)z\| \le \hat{\kappa}_g \|x - y\| \, \|z\| \quad \text{for all } x, y, z \in X. \tag{4.1.4}$$

From Assumption 4.1 we can use Taylor series to obtain an initial residual bound

$$\|w_{k+1}\| \le ((1 - \beta_{k-1}) + \kappa_g \beta_{k-1}) \|w_k^\alpha\| + \frac{\hat{\kappa}_g}{2} \sum_{n=k-m_{k-1}}^{k-1} (\|e_{n+1}\| + \|e_n\|) \sum_{j=n}^{k-1} \|e_j \gamma_j^k\| . \tag{4.1.5}$$

The details in obtaining this bound are given in Subsection 4.3.1. The majority of the work in obtaining our main results come from controlling the higher-order terms in (4.1.5) in terms of residuals rather than differences of consecutive iterates. To do so we will make additional assumptions on the nondegeneracy of the fixed-point operator and (for $m > 1$) on the angles between subspaces spanned by the columns of matrix F_k of Algorithm 1.1 at each iteration. Even before making those assumptions we see directly from (4.1.5) that for linear operators, the higher-order terms do not appear because $\widehat{\kappa}_g$ Assumption 4.1 is zero for linear operators g.

4.1.2 ▪ Nondegeneracy assumption

Our next assumption generalizes the requirement of a contractive fixed-point operator. The contractive assumption was used in the convergence analysis frameworks of [50, 77, 91, 120]. As discussed below, this more general assumption is automatically satisfied at each iteration for contractive fixed-point operators and is also locally satisfied for finding zeros of nondegenerate operators.

Assumption 4.2. *The stage-j iterates and residuals satisfy the relationship*

$$\|\delta w_{j+1}\| \geq \sigma \|e_j\|. \tag{4.1.6}$$

Two cases where Assumption 4.2 is naturally satisfied are as follows.

1. If g is a contractive operator, then its Lipschitz constant given by Assumption 4.1 satisfies $\kappa_g < 1$, and by the triangle inequality,

$$\|w_{j+1} - w_j\| \geq \|x_j - x_{j-1}\| - \|g(x_j) - g(x_{j-1})\| \geq (1 - \kappa_g) \|e_j\|.$$

   Then (4.1.6) is satisfied with $\sigma = (1 - \kappa_g)$.

2. For finding a zero of function f using the fixed-point operator $g(x) = f(x) + x$, the nonlinear residual is $w_{j+1} = g(x_j) - x_j = f(x_j)$. Assumption 4.2 is then satisfied locally if the smallest singular value of the Jacobian f' is uniformly bounded away from zero in the vicinity of a solution, and $\|x_j - x_{j-1}\|$ is small enough.

   Specifically, suppose for each x, y in some ball $B_* \subset X$ containing x_* a zero of f it holds that $\|f'(x)y\| \geq \sigma_f \|y\|$ for some $\sigma_f > 0$. This is similar to the usual assumption for Newton iterations that the Jacobian is nondegenerate at a solution. Then, under Assumption 4.1, and using (4.1.4), we have

$$\begin{aligned}
\|f(x) - f(y)\| &= \left\| f'(y)(x-y) + \int_0^1 (f'(y + t(x-y)) - f'(y))(x-y)\,\mathrm{d}t \right\| \\
&\geq \sigma_f \|x - y\| - \frac{\kappa_g}{2} \|x - y\|^2 \\
&\geq \frac{\sigma_f}{2} \|x - y\| \quad \text{for } \|x - y\| \leq \frac{\sigma_f}{\hat{\kappa}_g}.
\end{aligned}$$

   Thus for $\|e_j\| \leq \sigma_f / \hat{\kappa}_g$ it holds that $\|\delta w_{j+1}\| \geq \frac{\sigma_f}{2} \|e_j\|$, which satisfies (4.1.6) with $\sigma = \sigma_f / 2$.

   We would like to further point out that this case can be thought of as an opportunity to apply preconditioning to improve the constant σ by bounding the constant σ_f corresponding to a preconditioned operator f further away from zero.

If neither of the two cases above can be shown to hold, assumption (4.1.6) can also be enforced by monitoring $\|\delta w_{j+1}\|$ and $\|e_j\|$ upon the computation of each update step w_{k+1} and restarting the iteration if their ratio drops below a user-chosen threshhold σ_0. This restarting strategy is mentioned in [90], which studies AA applied to a regularized Bingham model; however, when the ratio was monitored in that study it was found to stay bounded well above zero, and restarting was not applied. These results are shown at the end of the numerical Section 4.5 of this chapter.

For algorithmic depth $m > 1$ we will make one more assumption. Similarly to Assumption 4.2, it can be *enforced* rather than *assumed*. The filtering strategy that enforces the next assumption is readily seen to improve convergence, particularly in the preasymptotic regime. Examples of this are shown in Section 4.5 of this chapter and again in Chapter 5.

4.1.3 ▪ Angle assumption

This third assumption requires that each subsequently older column in the matrix F_k of Algorithm 1.1 sufficiently contributes to expanding the span of the columns that lie to its left (those that contain newer information). It is written in terms of the directions sines between each column vector and the span of column vectors to its left. Posing it in this way makes it natural to monitor and enforce by reading off the diagonal entries of the matrix R of the QR decomposition of F_k. This is explained in detail in Section 4.3.3.

Assumption 4.3. *Let $f_1, f_2, \ldots, f_m$ be the columns of matrix F_k of Algorithm 1.1. Assume there is a constant $c_s > 0$ such that*

$$|\sin(f_j, \operatorname{span}\{f_1, f_2, \ldots, f_{j-1}\})| \geq c_s, \; j = 2, \ldots, m. \tag{4.1.7}$$

The satisfaction of Assumption 4.3 then implies there is a constant $0 \leq c_t < 1$ which satisfies

$$|\cos(f_j, f_n\})| \leq c_t, \; j = 2, \ldots, m, \quad \text{and} \quad n = 1, \ldots, j-1.$$

The statement of Assumption 4.3 is equivalent to the assumption in [88, Lemma 5.2], the content of which appears here as Lemma 4.14. The difference in appearance is the assumption in [88] being written in terms of the "global" indices of δw_j (denoted there as v_j), by which (4.1.7) states that

$$|\sin(\delta w_{n+1}, \operatorname{span}\{\delta w_{n+2}, \ldots, \delta w_{j+1}\})| \geq c_s, \; n = j - m + 1, \ldots, j-1.$$

The "local" indexing in (4.1.7) with respect to each matrix F_k seems to be easier to follow and is more practically applicable.

4.2 ▪ Main results

First we state the main results for depth $m = 1$. Both the result and its proof are simpler in form than the more general result, stated after. The main difference is that for $m = 1$ there is no need to address the independence of, or the angles between, the columns of matrix F_k of Algorithm 1.1, as it has only one column.

Theorem 4.4. *Let Assumption 4.1 hold, and let $m = 1$ in Algorithm 1.1. Assume there is a constant $\sigma > 0$ for which the residuals on stages $j = k-2, k-1, k$ satisfy Assumption 4.2. Then the following bound holds for the nonlinear residual $\|w_{k+1}\|$ generated by Algorithm 1.1 with depth $m = 1$:*

$$\begin{aligned}\|w_{k+1}\| \leq \|w_k\| \Bigg\{ &\theta_k\big((1-\beta_{k-1}) + \kappa_g \beta_{k-1}\big) + \frac{\hat{\kappa}_g}{2}\sigma^{-1}\sqrt{1-\theta_k^2} \\ \times \bigg(&\|w_k\| \left(\sigma^{-1}\sqrt{1-\theta_k^2} + \beta_{k-1}\theta_k\right) + \|w_{k-1}\| \left(\sigma^{-1}\sqrt{1-\theta_{k-1}^2} + \beta_{k-2}\theta_{k-1}\right)\bigg)\Bigg\}. \end{aligned} \tag{4.2.1}$$

This result shows how the first-order term is scaled by the optimization gain θ_k and also how the higher-order terms are scaled by $\sqrt{1-\theta_k^2}$. Thus, for θ_k close to 1, the contribution from the higher-order terms is necessarily close to zero, and the total residual bound approaches that of the original fixed-point iteration. For θ_k close to 0, the contribution from the first-order (linear) term is reduced. If the previous residuals are sufficiently small, we see an improved linear rate of convergence. If, however, the previous residuals that appear in the bound are not sufficiently less than 1, then the higher-order terms can substantially contribute to and even dominate the residual, potentially preventing convergence. This last scenario is more likely to be the case in the preasymptotic regime. In the next result, which holds for algorithmic depths $m \geq 1$, we see these properties continue to hold.

Theorem 4.5. *Let Assumption 4.1 hold. Assume there is a constant $\sigma > 0$ for which the residuals on stages $j = \min\{0, k - 2m\}, \ldots, k$, satisfy Assumption 4.2. Suppose Assumption 4.3 holds for $j = k$. Then the following bound holds for the nonlinear residual $\|w_{k+1}\|$ generated by Algorithm 1.1 with depth m:*

$$\|w_{k+1}\| \leq \|w_k\| \Bigg\{ \theta_k((1 - \beta_{k-1}) + \kappa_g \beta_{k-1}) + \frac{\hat{\kappa}_g}{2} \Bigg(\|w_k\| \, h(\theta_k) h_{k-1}(\theta_k) \\ + 2 \sum_{n=k-m_{k-1}+1}^{k-1} \Big(\|w_n\| \, h(\theta_n) \sum_{j=n}^{k-1} h_j(\theta_k) \Big) \\ + \left\|w_{k-m_{k-1}}\right\| h(\theta_{k-m_{k-1}}) \sum_{j=k-m_{k-1}}^{k-1} h_j(\theta_k) \Bigg) \Bigg\}, \tag{4.2.2}$$

where

$$h(\theta_j) = C_{F,j} \sqrt{1 - \theta_j^2} + \beta_{j-1} \theta_j, \qquad h_j(\theta_k) = C_{j,k} \sqrt{1 - \theta_k^2}, \tag{4.2.3}$$

and the constants $C_{F,j}$ and $C_{j,k}$ depend on the parameters m, σ, and c_s, satisfying the following properties:

- *$C_{F,j} \sim 1/\sigma$, $C_{F,j} \geq m_{j-1}/\sigma$, and $C_{F,j}$ is decreasing with respect to $c_s \in (0, 1]$ with $C_{F,j} = m_{j-1}/\sigma$ for $c_s = 1$ and $C_{F,j} \to \infty$ as $c_s \to 0$.*
- *$C_{j,k} \sim 1/\sigma$, $C_{j,k} \geq 1/\sigma$, and $C_{j,k}$ is decreasing with respect to $c_s \in (0, 1]$ with $C_{j,k} = 1/\sigma$ for $c_s = 1$ and $C_{j,k} \to \infty$ as $c_s \to 0$.*

The precise definitions of constants $C_{F,j}$ and $C_{j,k}$ can be found in the proof of Theorem 4.5 in Section 4.3.3 in (4.3.23) and (4.3.25), respectively.

The results above show the higher-order terms are bounded by

$$\mathcal{O}(\|w_k\|^2) + \mathcal{O}(\|w_k\| \, \|w_{k-1}\|) + \cdots + \mathcal{O}(\|w_k\| \, \left\|w_{k-m_{k-1}}\right\|),$$

where the constants in front of each term remain bounded so long as the constants σ, from Assumption 4.2, and c_s, from Assumption 4.3, are bounded away from zero.

In the case that $m = 1$, the general result (4.2.2) reduces to (4.2.1). To see this, first notice that with $m_{k-1} = m = 1$, the terms summed from $n = k - m_{k-1} + 1$ to $n = k - 1$ do not appear, and the only index that appears in the terminal sum of (4.2.2) is $j = k - 1$. Then we can refer to (4.3.23) and (4.3.25) to see that $C_{F,k}$, $C_{F,k-1}$, and $C_{k-1,k}$ each reduce to σ^{-1} for $m = 1$. Rearranging terms then yields (4.2.1). This reinforces the idea that the main theoretical difference between the main result for $m = 1$ and the main result for $m > 1$, beyond the increased number of terms corresponding to the increased algorithmic depth, is the effect of the lack of orthogonality of the columns of F_k in the higher-order terms of the residual bound.

Remark 4.6 (Removal of assumptions on the optimization coefficients). *Neither of Theorems 4.4 or 4.5 makes an assumption on the boundedness of the optimization coefficients. For the interested reader, the explanation can be found in the technical proofs of the theorems in section 4.3, according to the following outline.*

First, referring to the initial bound on $\|w_{k+1}\|$ via Taylor series in (4.1.5), *as derived in section 4.3.1, we see that each optimization coefficient γ_j^k appears only as a part of the normed product $\|e_j\gamma_j^k\|$. For $m = 1$, this quantity is controlled by relating the optimization coefficient to the optimization gain in* (4.3.14), *and applying the nondegeneracy Assumption 4.2. For $m > 1$, a QR decomposition is used to accomplish the analogous estimate to* (4.3.14) *in* (4.3.27)-(4.3.29). *Here the additional condition of Assumption 4.3 arises that the direction sines of the angles between each column of F_k and the subspace spanned by the columns to its left are bounded away from zero. This essentially says that each additional column of F_k sufficiently contributes to its range. The particular form of this condition allows it to be enforced through a simple filtering procedure.*

As noted in subsection 4.1.1, since $\hat{\kappa}_g$ represents the Lipschitz constant of g', if g is linear, then $\hat{\kappa}_g = 0$ and thus all of the higher-order terms on the right-hand sides of (4.2.1) and (4.2.2) vanish. The main technicalities of these results are relevant only to nonlinear operators.

4.2.1 ▪ Contractive operators

We next show corollaries to Theorems 4.4 and 4.5 with simplified residual bounds for contractive operators, together with conditions for monotonic decrease of the residual. These results feature tighter bounds on the higher-order terms than in [50] and without assumptions on the boundedness of the optimization coefficients.

Corollary 4.7. *Suppose the hypotheses of Theorem 4.4, and suppose the Lipschitz constant of g satisfies $\kappa_g < 1$. Then the following bound holds on the nonlinear residual $\|w_{k+1}\|$ generated by Algorithm 1.1 with $m = 1$ and $\beta_k = \beta = 1$:*

$$\|w_{k+1}\| \le \|w_k\| \left\{ \theta_k \kappa_g + \frac{\sqrt{2}\hat{\kappa}_g}{(1-\kappa_g)^2} \sqrt{1-\theta_k^2} \Big(\|w_k\| + \|w_{k-1}\| \Big) \right\}. \tag{4.2.4}$$

After the first two consecutive iterations $j = k-1, k$ where the following inequality is satisfied,

$$\|w_j\| + \|w_{j-1}\| < \frac{\sqrt{1-\kappa_g^2(1-\kappa_g)^2}}{\sqrt{2}\hat{\kappa}_g}, \tag{4.2.5}$$

monotonic decrease of the residual is ensured.

Proof. From (4.2.1) with $\beta_k = 1$ and $\sigma = (1-\kappa_g)$, the residual $\|w_{k+1}\|$ satisfies

$$\|w_{k+1}\| \le \|w_k\| \left\{ \theta_k \kappa_g + \frac{\hat{\kappa}_g \sqrt{1-\theta_k^2}}{(1-\kappa_g)^2} \Big(\|w_k\| \left(\sqrt{1-\theta_k^2} + \theta_k \right) \right.$$
$$\left. + \|w_{k-1}\| \left(\sqrt{1-\theta_{k-1}^2} + \theta_{k-1} \right) \Big) \right\}. \tag{4.2.6}$$

The maximum of $\sqrt{1-\theta^2}+\theta$ on $0 \le \theta \le 1$ is $\sqrt{2}$, attained at $\theta = 1/\sqrt{2}$. Applying this to θ_{k-1}, θ_k within the higher-order terms yields (4.2.4).

Following the same idea, maximizing the bracketed term on the right-hand side of (4.2.4) over θ_k,

$$\theta_k \kappa_g + \frac{\sqrt{2}\hat{\kappa}_g}{(1-\kappa_g)^2}\sqrt{1-\theta_k^2}\Big(\|w_k\| + \|w_{k-1}\| \Big) \le \sqrt{\kappa_g^2 + \frac{2\hat{\kappa}_g^2}{(1-\kappa_g)^4}\big(\|w_k\| + \|w_{k-1}\| \big)^2}.$$

Setting (the square of) the right-hand-side expression less than 1, it follows that $\|w_{k+1}\| < \|w_k\|$ under condition (4.2.5). If this condition is satisfied for two consecutive iterates, then $\|w_{k+1}\| < \|w_k\|$ and $\|w_k\| < \|w_{k-1}\|$, which is sufficient to ensure monotonic decrease of the sequence. □

This corollary quantifies (in the contractive setting) the transition from the preasymptotic regime where the residuals may be large to the asymptotic regime where the residuals are small enough that the higher-order terms "don't count," and previous convergence results such as those in [91] hold (see also [77, 120] for a different but related approach). This is next generalized in Corollary 4.8 for algorithmic depths $m > 1$ where it is sufficient for a similar condition to hold for $m+1$ consecutive iterates. While this monotonicity result holds only for contractive operators, this also includes *locally contractive* operators within their contractive regime.

Corollary 4.8. *Suppose the hypotheses of Theorem 4.5, and suppose the Lipschitz constant κ_g satisfies $\kappa_g < 1$. Then the following bound holds for the nonlinear residual $\|w_{k+1}\|$ generated by Algorithm 1.1 with $\beta_k = \beta = 1$:*

$$\|w_{k+1}\| \le \|w_k\| \left\{ \theta_k \kappa_g + \frac{C\sqrt{1+C^2}\sqrt{1-\theta_k^2}\hat{\kappa}_g}{2} \times \left(\|w_k\| + 2 \sum_{n=k-m_{k-1}+1}^{k-1} (k-n)\|w_n\| + m_{k-1}\|w_{k-m_{k-1}}\| \right) \right\}, \tag{4.2.7}$$

where

$$C = \max\left\{ \frac{1}{\sigma c_s}\left(\frac{c_t + c_s}{c_s}\right)^{m_k - 1}, C_{F,k+1} \right\}, \quad \text{with } \sigma = (1-\kappa_g). \tag{4.2.8}$$

After the first $m+1$ consecutive iterations $j = k-m, \ldots, k$ (assuming here for simplicity that $k \ge 2m$, so the subscript on m may be dropped) such that the following inequality is satisfied,

$$\|w_j\| + 2\sum_{n=j-m+1}^{k-1} \|w_n\| + \|w_{j-m}\| < \frac{2(1-\kappa_g)}{C\sqrt{1+C^2}\hat{\kappa}_g}, \tag{4.2.9}$$

monotonic decrease of the residual is assured.

The proof follows similarly to the $m = 1$ case in Corollary 4.7, with the additional steps of bounding the two types of h coefficients.

Proof. For each $\beta_j = 1$ and $\sigma = 1 - \kappa_g$, as in Assumption 4.2, the coefficients $h_n(\theta_k)$ are each bounded by $C\sqrt{1-\theta_k^2}$, with C given by (4.2.8). Applying this to (4.2.2) allows

$$\|w_{k+1}\| \leq \|w_k\| \left(\theta_k \kappa_g + \frac{C\hat{\kappa}_g\sqrt{1-\theta_k^2}}{2} \left(\|w_k\|\, h(\theta_k) \right.\right.$$
$$\left.\left. + 2 \sum_{n=k-m_{k-1}+1}^{k-1} (k-n)\, \|w_n\|\, h(\theta_n) + m_{k-1} \left\|w_{k-m_{k-1}}\right\| h(\theta_{k-m_{k-1}}) \right) \right). \quad (4.2.10)$$

The coefficients $h(\theta_j)$ are each bounded by $C\sqrt{1-\theta_j^2} + \theta_j \leq \sqrt{1+C^2}$. Applying this to (4.2.10) yields (4.2.7).

Maximizing in terms of θ_k, the square of the bracketed terms on the right-hand side of (4.2.7) is bounded by

$$\kappa_g^2 + \frac{C^2(1+C^2)\hat{\kappa}_g^2}{4} \left(\|w_k\| + 2 \sum_{n=k-m+1}^{k-1} (k-n)\, \|w_n\| + m\, \|w_{k-m}\| \right)^2. \quad (4.2.11)$$

Setting (4.2.11) less than 1 implies $\|w_{k+1}\| < \|w_k\|$ under the condition (4.2.9). Satisfaction of $\|w_{j+1}\| < \|w_j\|$ for $m+1$ consecutive iterates $j = k-m, \ldots, k$ then implies reduction in every subsequent residual. □

4.3 ▪ Technical proofs of theorems

We now delve into the technical proofs behind the main results. We will start with the expansion of the residual by Taylor series. The sections that follow show for $m = 1$ and then for $m \geq 1$ how to trade in the differences between consecutive iterates for residual terms so that the final residual bound is in terms of residuals only.

4.3.1 ▪ Expansion of the residual

Starting with the definition of the residual by (1.2.1) and expanding the iterate x_k by the update (1.2.3), the nonlinear residual w_{k+1} can be written as

$$w_{k+1} = g(x_k) - x_k = (g(x_k) - x_{k-1}^{\alpha}) - \beta_{k-1} w_k^{\alpha}. \quad (4.3.1)$$

The first term on the right-hand side of (4.3.1) can be expanded by (1.2.5). Using the property that the sum $\alpha_{k-1-m_{k-1}}^k + \cdots + \alpha_{k-2}^k + \alpha_{k-1}^k = 1$, then collecting the sums of the α_n^k coefficients into γ_j^k by

$$\gamma_j^k = \sum_{n=k-m_{k-1}-1}^{j-1} \alpha_n^k,$$

as in (1.2.6), we have

$$
\begin{aligned}
g(x_k) - x_{k-1}^{\alpha} &= \sum_{j=k-m_{k-1}-1}^{k-1} \alpha_j^k (g(x_k) - x_j) \\
&= \sum_{j=k-m_{k-1}-1}^{k-1} \alpha_j^k (g(x_j) - x_j) + \sum_{j=k-m_{k-1}}^{k} \left(\sum_{n=k-m_{k-1}-1}^{j-1} \alpha_n^k \right) (g(x_j) - g(x_{j-1})) \\
&= w_k^{\alpha} + \sum_{j=k-m_{k-1}}^{k} \gamma_j^k (g(x_j) - g(x_{j-1})).
\end{aligned} \tag{4.3.2}
$$

This equality shows the approximation to the fixed-point $g(x_k)$ is decomposed into the average of the previous iterates x_{k-1}^{α}, the average over previous updates w_k^{α} corresponding to the optimization problem from the last step, and a weighted sum over the differences of consecutive approximations. Due to Assumption 4.1, each term $g(x_j) - g(x_{j-1})$ has a Taylor expansion $g(x_j) - g(x_{j-1}) = \int_0^1 g'(z_j(t))e_j \,\mathrm{d}t$, where $z_j(t) = x_{j-1} + te_j$. Rewriting (4.3.1) with (4.3.2) with this expansion yields

$$
w_{k+1} = (1 - \beta_{k-1}) w_k^{\alpha} + \sum_{j=k-m_{k-1}}^{k} \gamma_j^k \int_0^1 g'(z_j(t)) e_j \,\mathrm{d}t. \tag{4.3.3}
$$

Adding and subtracting consecutive averages, each summand of the last term of (4.3.3) can be written as

$$
\int_0^1 g'(z_j(t))e_j \,\mathrm{d}t = \int_0^1 g'(z_k(t))e_j \,\mathrm{d}t + \sum_{n=j}^{k-1} \int_0^1 g'(z_n(t))e_j - g'(z_{n+1}(t))e_j \,\mathrm{d}t. \tag{4.3.4}
$$

Summing over the j's, the sum on the right-hand side of (4.3.3) may be expressed as

$$
\begin{aligned}
\sum_{j=k-m_{k-1}}^{k} \gamma_j^k \int_0^1 g'(z_j(t))e_j \,\mathrm{d}t = &\int_0^1 g'(z_k(t)) \sum_{j=k-m_{k-1}}^{k} \gamma_j^k e_j \,\mathrm{d}t \\
&+ \sum_{j=k-m_{k-1}}^{k-1} \sum_{n=j}^{k-1} \gamma_j^k \int_0^1 g'(z_n(t))e_j - g'(z_{n+1}(t))e_j \,\mathrm{d}t.
\end{aligned} \tag{4.3.5}
$$

To simplify the first sum in (4.3.5), first note that $\gamma_k^k = 1$; then apply (1.2.4) to the remaining terms with index k shifted to $k-1$, followed by (1.2.3), to obtain

$$
\sum_{j=k-m_{k-1}}^{k} \gamma_j^k e_j = e_k + E_{k-1}\gamma^k = x_k - x_{k-1}^{\alpha} = \beta_{k-1} w_k^{\alpha}. \tag{4.3.6}
$$

Putting (4.3.6) together with (4.3.5) and (4.3.3) then yields

$$
\begin{aligned}
w_{k+1} = &\int_0^1 (1 - \beta_{k-1}) w_k^{\alpha} + \beta_{k-1} g'(z_k(t)) w_k^{\alpha} \,\mathrm{d}t \\
&+ \sum_{j=k-m_{k-1}}^{k-1} \sum_{n=j}^{k-1} \int_0^1 (g'(z_n(t)) - g'(z_{n+1}(t))) e_j \gamma_j^k \,\mathrm{d}t.
\end{aligned} \tag{4.3.7}
$$

The terms summed over are next bounded using Assumption 4.1. For nonlinear operators, the higher-order terms are scaled by $\hat{\kappa}_g > 0$, the Lipschitz constant of g'. Intuitively, this expansion leads to a local result, as when the difference between iterates is sufficiently small, the graph of a C^1 Lipschitz function at or between those iterates is nearly linear.

Taking norms in (4.3.7) and applying Assumption 4.1 followed by triangle inequalities applied to the terms of the final sum produces the expansion of $\|w_{k+1}\|$ in terms of $\|w_k^\alpha\|$ and $\|e_j\|$, $j = k - m_k, \ldots, k$, by

$$\begin{aligned}\|w_{k+1}\| &\le ((1-\beta_{k-1}) + \kappa_g\beta_{k-1})\|w_k^\alpha\| + \frac{\hat{\kappa}_g}{2}\sum_{j=k-m_{k-1}}^{k-1}\|e_j\gamma_j^k\|\sum_{n=j}^{k-1}(\|e_{n+1}\| + \|e_n\|)\\ &= ((1-\beta_{k-1}) + \kappa_g\beta_{k-1})\|w_k^\alpha\| + \frac{\hat{\kappa}_g}{2}\sum_{n=k-m_{k-1}}^{k-1}(\|e_{n+1}\| + \|e_n\|)\sum_{j=n}^{k-1}\|e_j\gamma_j^k\|,\end{aligned} \tag{4.3.8}$$

where the last equality follows from reindexing the sums. The next step is to bound the $\|e_j\|$ terms by $\|w_j\|$ terms. This will be shown first in the simpler case of depth $m = 1$ in Section 4.3.2 and then extended to more general depths $m > 1$ in Section 4.3.3.

4.3.2 ▪ Acceleration for depth $m = 1$

For depth $m = 1$, the matrix F_k has only one column, which removes some technicalities from the analysis, such as the general lack of orthogonality between the columns of F_k when $m > 1$. It is useful to use this case to overview the general framework and to introduce the extension to a noncontractive setting.

Lemma 4.9. *Let Assumption 4.1 hold, and let $m = 1$ in Algorithm 1.1. Assume there is a constant $\sigma > 0$ for which the residuals on stages $j+1$ and j satisfy Assumption 4.2. Then the following bound holds on the difference between consecutive accelerated iterates:*

$$\|e_{j+1}\| \le \|w_{j+1}\|\left(\sigma^{-1}\sqrt{1-\theta_{j+1}^2} + \beta_j\theta_{j+1}\right). \tag{4.3.9}$$

Proof. The update (1.2.3) for the case $m = 1$ is

$$x_{j+1} = (1-\gamma_j^{j+1})x_j + \gamma_j^{j+1}(x_{j-1}) + \beta_j w_{j+1}^\alpha. \tag{4.3.10}$$

Taking norms and applying (1.2.7) allows

$$\|e_{j+1}\| = \|x_{j+1} - x_j\| \le |\gamma_j^{j+1}|\,\|e_j\| + \beta_j\theta_{j+1}\|w_{j+1}\|. \tag{4.3.11}$$

Inequality (4.3.11) will be used to trade terms of the form $\|e_{j+1}\|$ for expressions in terms of $\|w_{j+1}\|$. The argument follows by relating the optimization coefficient γ_j^{j+1} to the optimization gain θ_{j+1} and applying Assumption 4.2.

For $m = 1$, the coefficient γ_j^{j+1} can be explicitly written as

$$\gamma_j^{j+1} = \frac{(w_{j+1}, \delta w_{j+1})}{\|\delta w_{j+1}\|^2}. \tag{4.3.12}$$

In particular, this determines the decomposition of w_{j+1} into $w_R = \gamma_j^{j+1}\delta w_{j+1}$, in the range of δw_{j+1}, and $w_N = w_{j+1}^\alpha = w_{j+1} - \gamma_j^{j+1}\delta w_{j+1}$ in the null space of $(\delta w_{j+1})^T$. By the orthogonality of w_R and w_N it follows that

$$\|w_{j+1}\|^2 = \|w_R\|^2 + \|w_N\|^2 = \left\|\gamma_j^{j+1}\delta w_{j+1}\right\|^2 + \left\|w_{j+1}^\alpha\right\|^2$$
$$= (\gamma_j^{j+1})^2 \|\delta w_{j+1}\|^2 + \theta_{j+1}^2 \|w_{j+1}\|^2, \tag{4.3.13}$$

by which

$$|\gamma_j^{j+1}| = \sqrt{1-\theta_{j+1}^2}\frac{\|w_{j+1}\|}{\|\delta w_{j+1}\|} \text{ and } \theta_{j+1} = \sqrt{1 - \frac{(w_{j+1}, \delta w_{j+1})^2}{\|w_{j+1}\|^2 \|\delta w_{j+1}\|^2}}. \tag{4.3.14}$$

The expression for θ_{j+1} in (4.3.14) can be recognized as the (absolute value of the) direction sine between w_{j+1} and δw_{j+1}, by which the expression for γ_j^{j+1} is the direction cosine, scaled by $\|w_{j+1}\| / \|\delta w_{j+1}\|$. Applying the expression for γ_j^{j+1} in (4.3.14) to (4.3.11) yields

$$\|e_{j+1}\| \le \sqrt{1-\theta_{j+1}^2}\frac{\|w_{j+1}\|}{\|\delta w_{j+1}\|} \|e_j\| + \beta_j \theta_{j+1} \|w_{j+1}\|. \tag{4.3.15}$$

Applying now the key inequality (4.1.6) to (4.3.15) yields

$$\|e_{j+1}\| \le \sigma^{-1}\sqrt{1-\theta_{j+1}^2}\, \|w_{j+1}\| + \beta_j \theta_{j+1} \|w_{j+1}\|, \tag{4.3.16}$$

establishing the result (4.3.9) □

Remark 4.10. *In the second enumerated case of Assumption 4.2 where f' is nondegenerate, the results of Lemma 4.9 and $0 < \beta_j \le 1$ show*

$$\|e_{j+1}\| \le \left(\frac{2}{\sigma_f}\sqrt{1-\theta_{j+1}^2} + \theta_{j+1}\right) \|w_{j+1}\| \le \sqrt{1+4/\sigma_f^2}\, \|w_{j+1}\|,$$

where the last bound was obtained by maximizing the previous expression with respect to θ_{j+1}. Setting this expression no greater than $\sigma_f/\hat{\kappa}_g$ it follows that $\|w_{j+1}\| \le \sigma_f^2/(\hat{\kappa}_g\sqrt{\sigma_f^2+4})$ is sufficient to ensure $\|e_{j+1}\| \le \sigma_f/\hat{\kappa}_g$, which implies satisfaction of Assumption 4.2 on the subsequent iteration.

Relation (4.3.16) is now used in the expansion of the residual (4.3.8) to bound $\|w_{k+1}\|$. Now we are ready to prove Theorem 4.4.

Proof. Expanding the residual by (4.3.8) yields for depth $m = 1$,

$$\|w_{k+1}\| \le \theta_k\big((1-\beta_{k-1}) + \kappa_g \beta_{k-1}\big) \|w_k\| + \frac{\hat{\kappa}_g}{2} (\|e_k\| + \|e_{k-1}\|) |\gamma_{k-1}^k| \|e_{k-1}\|.$$

Applying (4.3.16) with $j = k-1$ and $j = k-2$ then allows

$$\|w_{k+1}\| \le \theta_k\big((1-\beta_{k-1}) + \kappa_g \beta_{k-1}\big) \|w_k\| + \frac{\hat{\kappa}_g}{2}\Bigg(\|w_k\| \left(\sigma^{-1}\sqrt{1-\theta_k^2} + \beta_{k-1}\theta_k\right)$$
$$+ \|w_{k-1}\| \left(\sigma^{-1}\sqrt{1-\theta_{k-1}^2} + \beta_{k-2}\theta_{k-1}\right)\Bigg) |\gamma_{k-1}^k| \|e_{k-1}\|. \tag{4.3.17}$$

For the last term on the right-hand side of (4.3.17), applying relation (4.3.14) with hypothesis (4.1.6) yields

$$|\gamma_{k-1}^k| \|e_{k-1}\| \le \sigma^{-1}\sqrt{1-\theta_k^2}\, \|w_k\|,$$

by which

$$\|w_{k+1}\| \le \theta_k\big((1-\beta_{k-1})+\kappa_g\beta_{k-1}\big)\|w_k\| + \frac{\hat{\kappa}_g}{2}\Bigg(\|w_k\|\left(\sigma^{-1}\sqrt{1-\theta_k^2}+\beta_{k-1}\theta_k\right)$$
$$+\|w_{k-1}\|\left(\sigma^{-1}\sqrt{1-\theta_{k-1}^2}+\beta_{k-2}\theta_{k-1}\right)\Bigg)\sigma^{-1}\sqrt{1-\theta_k^2}\,\|w_k\|, \qquad (4.3.18)$$

establishing the result (4.2.1). □

4.3.3 ▪ Acceleration for depth $m > 1$

The analysis for $m > 1$ is more complicated than for $m = 1$, if only because in the optimization problem for $m = 1$, the matrix F_k has only one column. For $m > 1$, the columns of F_k are in general not orthogonal, and the estimates that follow show how detrimental this lack of orthogonality can be to the convergence rate. First some standard results from numerical linear algebra are recalled. Then Theorem 4.4 is generalized to $m > 1$.

Proposition 4.11. *Let R_j be a $j \times j$ upper triangular matrix given by*

$$R_j = \begin{pmatrix} R_{j-1} & b_j \\ 0 & r_{jj} \end{pmatrix},$$

where R_{j-1} is an invertible $j-1 \times j-1$ upper triangular matrix, b_j is a $j-1 \times 1$ vector of values, and $r_{jj} \neq 0$. Then R_j is invertible and the inverse matrix satisfies

$$R_j^{-1} = \begin{pmatrix} R_{j-1}^{-1} & c_j \\ 0 & r_{jj}^{-1} \end{pmatrix}.$$

The next two results are specific to the economy (or thin) QR decomposition of $n \times m$ matrix A (see, for instance, [65, Chapter 5]). The following notation will be used throughout the remainder of this section. For $u, v \in \mathbb{R}^n$, let $\cos(u,v) = (u,v)/(\|u\|\,\|v\|)$ be the usual direction cosine between vectors u and v, with the corresponding direction sine satisfying $\sin^2(u,v) = 1-\cos^2(u,v)$. Let $\mathcal{A}_j$ be the subspace of $\mathbb{R}^n$ given by $\mathcal{A}_j = \operatorname{span}\{a_1,\ldots,a_j\}$, with orthogonal basis $\{q_1,\ldots,q_j\}$; then let $\sin^2(u,\mathcal{A}_j) = 1-\sum_{i=1}^{j}\cos^2(u,q_i)$ denote the square of the direction sine between vector u and subspace $\mathcal{A}_j$.

Proposition 4.12. *Let $\hat{Q}\hat{R}$ be the economy QR decomposition of a matrix $A \in \mathbb{R}^{n\times m}$, $n \ge m$, where A has columns $a_1,\ldots,a_m$, and $\hat{Q}$ has orthonormal columns $q_1,\ldots,q_m$. Then*

$$r_{jj}^2 = \|a_j\|^2\sin^2(a_j,\mathcal{A}_{j-1}),\; j = 1,\ldots,m. \qquad (4.3.19)$$

The proof is standard and follows from writing the columns of $\hat{Q}$ as $q_j = v_j/\|v_j\|$ with $v_j = a_j - \sum_{i=1}^{j-1}(q_i,a_j)q_i$. Then $r_{jj}^2 = \|v_j\|^2 = \|a_j\|^2 - \sum_{i=1}^{j-1}(q_i,a_j)^2$ from orthogonality. Factoring out $\|a_j\|^2$ from each term yields the result.

The next technical lemma gives a bound on the elements of $\hat{R}^{-1}$.

Lemma 4.13. *Let $\hat{Q}\hat{R}$ be the economy QR decomposition of matrix $A \in \mathbb{R}^{n\times m}$, $n \ge m$, where A has columns $a_1,\ldots,a_m$, $\hat{Q}$ has orthonormal columns $q_1,\ldots,q_m$, and $\hat{R} = (r_{ij})$ is an invertible upper-triangular $m \times m$ matrix. Let $\hat{R}^{-1} = (s_{ij})$.*

Suppose there is a constant $0 < c_s \le 1$ *such that* $|\sin(a_j, \mathcal{A}_{j-1})| \ge c_s,\ j = 2, \dots, m$, *which implies another constant* $0 \le c_t < 1$ *with* $|\cos(a_j, q_i)| \le c_t,\ j = 2, \dots, m$ *and* $i = 1, \dots, j-1$. *Then it holds that*

$$s_{11} = \frac{1}{\|a_1\|}, \qquad s_{ii} \le \frac{1}{\|a_i\|\, c_s},\ i = 2, \dots, m, \tag{4.3.20}$$

$$|s_{1j}| \le \frac{c_t(c_t + c_s)^{j-2}}{\|a_1\|\, c_s^{j-1}}, \qquad |s_{ij}| \le \frac{c_t(c_t + c_s)^{j-i-1}}{\|a_i\|\, c_s^{j-i+1}} \tag{4.3.21}$$

for $i = 2, \dots, m-1$ *and* $j = i+1, \dots, m$.

The constant $c_s > 0$ ensures the full rank of A and essentially bounds $\hat{Q}$ away from degeneracy, assuring sufficient linear independence of its columns. The results reflect that if the columns of A are close to (or actually) orthogonal, then c_t and the off-diagonal elements are close to (or actually) zero. The proof of Lemma 4.13 is deferred until the end of this chapter.

The next lemma generalizes Lemma 4.9 to $m > 1$. The technical difficulty of the more complicated relationship between the optimization coefficients and optimization gain is handled by expressing both in terms of a QR decomposition and then making use of Lemma 4.13.

Lemma 4.14. *Let Assumption 4.1 hold, let Assumption 4.2 hold with constant* σ *for* $n = j - m, \dots, j$, *and let Assumption 4.3 hold for* $k = j$. *Then the following bound holds for the difference between consecutive iterates* $e_{j+1} = x_{j+1} - x_j$*:*

$$\|e_{j+1}\| \le \|w_{j+1}\| \left(C_{F,j+1}\sqrt{1 - \theta_{j+1}^2} + \beta_j \theta_{j+1} \right), \tag{4.3.22}$$

where the constant $C_{F,j+1}$ *is given by*

$$C_{F,j+1} := \sigma^{-1} \left(1 + \frac{(1 + c_t)\sum_{l=1}^{m_j - 1} \binom{m_j - 1}{l} c_t^{l-1} c_s^{m_j - l - 1}}{c_s^{m_j - 1}} \right). \tag{4.3.23}$$

Additionally, the following bounds hold for terms of the form $\|e_n \gamma_n^{j+1}\|$*:*

$$\|e_n \gamma_n^{j+1}\| \le C_{n,j+1}\sqrt{1 - \theta_{j+1}^2}\, \|w_{j+1}\|,\ n = j - m_j, \dots, j, \tag{4.3.24}$$

where the constants $C_{n,j+1}$ *are given by*

$$C_{n,j+1} := \begin{cases} \frac{1}{\sigma}\left(\frac{c_t + c_s}{c_s}\right)^{m_j - 1}, & n = j, \\ \frac{1}{\sigma c_s}\left(\frac{c_t + c_s}{c_s}\right)^{m_j - (j - n + 1)}, & n = j - m_j, \dots, j - 1. \end{cases} \tag{4.3.25}$$

The additional assumption of (4.1.7) not found in the $m = 1$ case requires that the columns of the matrix F_k used in the least-squares problem (4.1.1), indexed locally as $f_1, f_2, \dots, f_m$, or indexed globally as $\delta w_{j+1}, \dots, \delta w_{j-m+2}$, maintain sufficient linear independence. We show in Section 4.4 of this chapter how this condition can be easily enforced, and we show in Chapter 5 how to generalize this strategy to ensure the least-squares problem solved at each iteration is well conditioned.

Proof. Throughout this proof, depth m_j will be denoted by m for simplicity. Starting with the update for x_{j+1} from (1.2.3) and (1.2.4), defined for optimization coefficients γ^{j+1} from (4.1.1), and the matrix E_j given by (4.1.3), shows $x_{j+1} - x_j = -E_j\gamma^{j+1} + \beta_k w_{j+1}^\alpha$. Taking norms and applying (1.2.7) yields

$$\|e_{j+1}\| \leq \left\|E_j\gamma^{j+1}\right\| + \beta_j\theta_{j+1}\left\|w_{j+1}\right\|. \tag{4.3.26}$$

By (4.1.1), the coefficients γ^{j+1} are the least-squares solution to $F_j\gamma^{j+1} = w_{j+1}$, where F_j is given by (4.1.3). Using an economy QR decomposition then provides $\hat{R}\gamma^{j+1} = \hat{Q}^T w_{j+1}$, by which the bound (4.3.26) may be written as

$$\|e_{j+1}\| \leq \left\|E_j\hat{R}^{-1}\hat{Q}^T w_{j+1}\right\| + \beta_j\theta_{j+1}\left\|w_{j+1}\right\|. \tag{4.3.27}$$

The first term on the right-hand side of (4.3.27) can be bounded in terms of $\|w_{j+1}\|$ by considering an explicit expression for the optimization gain θ_{j+1}, as first discussed in [50]. From (1.2.7) and the unique decomposition $w_{j+1} = w_R + w_N$ with $w_R \in \operatorname{Range}(F_j)$ and $w_N \in \operatorname{Null}((F_j)^T)$, the null-space component w_N is the least-squares residual satisfying

$$\|w_N\| = \left\|F_j\gamma^{j+1} - w_{j+1}\right\| = \left\|w_{j+1}^\alpha\right\| = \theta_{j+1}\left\|w_{j+1}\right\|,$$

meaning $\theta_{j+1} = \sqrt{1 - \left\|\hat{Q}^T w_{j+1}\right\|^2 / \left\|w_{j+1}\right\|^2}$ or, by rearranging,

$$\left\|w_{j+1}\right\|\sqrt{1-\theta_{j+1}^2} = \left\|\hat{Q}^T w_{j+1}\right\|. \tag{4.3.28}$$

The first term on the right-hand side of (4.3.27) can now be controlled by (4.3.28), yielding

$$\left\|E_j\hat{R}^{-1}\hat{Q}^T w_{j+1}\right\| \leq \left\|E_j\hat{R}^{-1}\right\|\left\|\hat{Q}^T w_{j+1}\right\| \leq \left\|E_j\hat{R}^{-1}\right\|\left\|w_{j+1}\right\|\sqrt{1-\theta_{j+1}^2}. \tag{4.3.29}$$

It remains to bound $\left\|E_j\hat{R}^{-1}\right\|$. Notice that in the case $m = 1$ this is simply $\|e_j\| / \|\delta w_{j+1}\|$, which is bounded by σ^{-1} via Assumption 4.2. For $m > 1$, we expect to still have a factor of σ^{-1} in our bound, but also to have a factor that takes into account the general lack of orthogonality between the columns of F_k, which is encoded in the matrix $\hat{R}$.

We proceed by writing out $E_j\hat{R}^{-1}$ column by column and bounding the matrix 2-norm of $E_j\hat{R}^{-1}$ by the vector 2-norms of its columns. Using triangle inequalities to separate terms multiplied by different columns of E_j, we can them regroup them, and apply the results of Lemma 4.13, the hypotheses of which are satisfied by Assumption 4.3. Writing $\hat{R}^{-1} = S = (s_{ij})$,

$$\begin{aligned}
\left\|E_j\hat{R}^{-1}\right\| &= \left\|\begin{pmatrix} e_j s_{11} & e_j s_{12} + e_{j-1}s_{22} & \dots & \sum_{n=1}^m e_{j-i+1}s_{nm}\end{pmatrix}\right\| \\
&\leq \left(\left\|e_j s_{11}\right\|^2 + \left\|e_j s_{12} + e_{j-1}s_{22}\right\|^2 + \dots + \left\|\sum_{n=1}^m e_{j-n+1}s_{nm}\right\|^2\right)^{1/2} \\
&\leq \|e_j\|\,|s_{11}| + (\|e_j\|\,|s_{12}| + \|e_{j-1}\|\,|s_{22}|) + \dots + \left(\sum_{n=1}^m \|e_{j-n+1}\|\,|s_{nm}|\right),
\end{aligned} \tag{4.3.30}$$

where the first inequality follows from the standard bounds of the matrix 2-norm by the Frobenius norm, and the last follows from triangle inequalities. Regrouping the terms in (4.3.30) by each column of E_j yields

$$\left\|E_j\hat{R}^{-1}\right\| \leq \|e_j\| \sum_{n=1}^{m} |s_{1n}| + \|e_{j-1}\| \left|\sum_{n=2}^{m} |s_{2n}| + \cdots + \|e_{j-m+1}\| \, |s_{mm}|. \tag{4.3.31}$$

The regrouping is not actually necessary, but it does make the indexing easier, as we next apply the results of the technical Lemma 4.13 and use finite geometric sums to add the contributions from each term.

For the first sum in (4.3.30), applying (4.3.20)-(4.3.21) of Lemma 4.13 and then taking the finite geometric sum and applying (4.1.6) of Assumption 4.2 produces the bound

$$\|e_j\| \sum_{n=1}^{m} |s_{1n}| \leq \frac{\|e_j\|}{\|\delta w_{j+1}\|} \left(1 + \sum_{n=2}^{m} \frac{c_t(c_t + c_s)^{n-2}}{c_s^{n-1}}\right) \leq \sigma^{-1} \left(\frac{c_t + c_s}{c_s}\right)^{m-1}. \tag{4.3.32}$$

Proceed similarly for the remaining sums of (4.3.30), indexed by $p = 2, \ldots, m$ (where $n = p$ is the starting index of each sum). Noting the additional factor of c_s in the denominator, we get

$$\|e_{j-p+1}\| \sum_{n=p}^{m} |s_{pn}| \leq \frac{\|e_{j-p+1}\|}{\|\delta w_{j-p+2}\| \, c_s} \left(1 + \sum_{n=p+1}^{m} \frac{c_t(c_t + c_s)^{n-(p+1)}}{c_s^{n-p}}\right) \leq \frac{1}{\sigma c_s} \left(\frac{c_t + c_s}{c_s}\right)^{m-p}. \tag{4.3.33}$$

Finally, adding the contributions from $p = 1$ to $p = 2, \ldots, m$ from (4.3.32) and (4.3.33) and applying the total to (4.3.30) yields, assuming $c_t \neq 0$,

$$\begin{aligned} \left\|E_j\hat{R}^{-1}\right\| &\leq \sigma^{-1} \left(\frac{(c_t + c_s)^{m-1}(c_t + 1) - c_s^{m-1}}{c_s^{m-1} c_t}\right) \\ &= \sigma^{-1} \left(1 + \frac{(1 + c_t) \sum_{j=1}^{m-1} \binom{m-1}{j} c_t^{j-1} c_s^{m-j-1}}{c_s^{m-1}}\right). \end{aligned} \tag{4.3.34}$$

If it so happens that $c_t = 0$, meaning the columns of F_k are orthogonal, then $c_s = 1$ and (4.3.34) is in agreement with summing the terms directly from (4.3.32) and (4.3.33), yielding $\left\|E_j\hat{R}^{-1}\right\| \leq m/\sigma$ in agreement in (4.3.34). Putting (4.3.34) together with (4.3.28) yields

$$\|e_{j+1}\| \leq C_{F,j+1}\sqrt{1 - \theta_{j+1}^2} \, \|w_{j+1}\| + \beta_j \theta_{j+1} \|w_{j+1}\|,$$

with $C_{F,j+1}$ given by (4.3.23) and hence the result (4.3.22).

For the second result, (4.3.24), expanding (4.3.27) shows

$$\begin{pmatrix} e_j \gamma_j^{j+1} & e_{j-1}\gamma_{j-1}^{j+1} & \cdots & e_{j-m+1}\gamma_{j-m+1}^{j+1} \end{pmatrix} = E_j \gamma^{j+1} = E_j \hat{R}^{-1} \hat{Q}^T w_{j+1}. \tag{4.3.35}$$

Accordingly, $e_{j-p+1}\gamma_{j-p+1}^{j+1} = e_{j-p+1} s^p \hat{Q}^T w_{j+1}$, where s^p is row p of $\hat{R}^{-1}$. Hence following (4.3.32) and applying (4.3.28) produces for the first column of (4.3.35)

$$\left\|e_j \gamma_j^{j+1}\right\| \leq \left\|e_j \sum_{n=1}^{m} s_{1n}\right\| \left\|w_{j+1}^{\alpha}\right\| \leq \sigma^{-1} \left(\frac{c_t + c_s}{c_s}\right)^{m-1} \sqrt{1 - \theta_{j+1}^2} \, \|w_{j+1}\|.$$

For the remaining columns, following now (4.3.33) allows

$$\left\|e_{j-p+1}\gamma_{j-p+1}^{j+1}\right\| \leq \left\|e_{j-p+1} \sum_{n=1}^{m} s_{pn}\right\| \left\|w_{j+1}^{\alpha}\right\| \leq \frac{1}{\sigma c_s} \left(\frac{c_t + c_s}{c_s}\right)^{m_j - p} \sqrt{1 - \theta_{j+1}^2} \, \|w_{j+1}\|,$$

which establishes the second result (4.3.24) with $n = j - p + 1$. □

Lemma 4.14 is now used to establish one-step residual bounds for general depths m. The proof of Theorem 4.5 follows the same essential outline as Theorem 4.4.

Proof. The expansion of the residual (4.3.8) from Section 4.3.1 shows

$$\|w_{k+1}\| \le ((1-\beta_{k-1}) + \kappa_g \beta_{k-1}) \|w_k^\alpha\| + \frac{\hat{\kappa}_g}{2} \sum_{n=k-m_{k-1}}^{k-1} (\|e_{n+1}\| + \|e_n\|) \sum_{j=n}^{k-1} \|e_j \gamma_j^k\| . \tag{4.3.36}$$

Opening up the first sum of (4.3.36) allows

$$\begin{aligned}
&\sum_{n=k-m_{k-1}}^{k-1} (\|e_{n+1}\| + \|e_n\|) \sum_{j=n}^{k-1} \|e_j \gamma_j^k\| \\
&= \|e_{k-m_{k-1}}\| \sum_{j=k-m_{k-1}}^{k-1} \|e_j \gamma_j^k\| + 2 \sum_{n=k-m_{k-1}+1}^{k-1} \|e_n\| \sum_{j=n}^{k-1} \|e_j \gamma_j^k\| + \|e_k\| \|e_{k-1} \gamma_{k-1}^k\| .
\end{aligned} \tag{4.3.37}$$

Applying now (4.3.22) then (4.3.24), the above (4.3.37) is bounded by

$$\begin{aligned}
&\|w_k\| h(\theta_k) \|e_{k-1}\gamma_{k-1}^k\| \\
&+2 \sum_{n=k-m_{k-1}+1}^{k-1} \left(\|w_n\| h(\theta_n) \sum_{j=n}^{k-1} \|e_j \gamma_j^k\| \right) + \|w_{k-m_{k-1}}\| h(\theta_{k-m_{k-1}}) \sum_{j=k-m_{k-1}}^{k-1} \|e_j \gamma_j^k\| \\
&\le \|w_k\| h(\theta_k) h_{k-1}(\theta_k) \|w_k\| + 2 \sum_{n=k-m_{k-1}+1}^{k-1} \left(\|w_n\| h(\theta_n) \sum_{j=n}^{k-1} h_j(\theta_k) \|w_k\| \right) \\
&+ \|w_{k-m_{k-1}}\| h(\theta_{k-m_{k-1}}) \sum_{j=k-m_{k-1}}^{k-1} h_j(\theta_k) \|w_k\| ,
\end{aligned} \tag{4.3.38}$$

where the terms $h(\theta_j)$ and $h_j(\theta_k)$ are given in (4.2.3) as

$$h(\theta_j) = C_{F,j}\sqrt{1-\theta_j^2} + \beta_{j-1}\theta_j \quad \text{and} \quad h_j(\theta_k) = C_{j,k}\sqrt{1-\theta_k^2},$$

with $C_{F,j}$ given by (4.3.23) and $C_{j,k}$ given by (4.3.25). Putting the bound of (4.3.38) back into (4.3.37) then yields

$$\begin{aligned}
\|w_{k+1}\| \le \|w_k\| \Bigg(&\theta_k((1-\beta_{k-1}) + \kappa_g \beta_{k-1}) \\
&+ \frac{\hat{\kappa}_g}{2} \Bigg(\|w_k\| h(\theta_k) h_{k-1}(\theta_k) + 2 \sum_{n=k-m_{k-1}+1}^{k-1} \left(\|w_n\| h(\theta_n) \sum_{j=n}^{k-1} h_j(\theta_k) \right) \\
&+ \|w_{k-m_{k-1}}\| h(\theta_{k-m_{k-1}}) \sum_{j=k-m_{k-1}}^{k-1} h_j(\theta_k) \Bigg) \Bigg),
\end{aligned}$$

the main result of Theorem 4.5. The asymptotics of $C_{F,j}$ and $C_{j,k}$ can be deduced directly from their definitions in (4.3.23) and (4.3.25). □

4.4 ▪ Practical guidance based on the theory

The results of Theorems 4.4-4.5 and Corollaries 4.7-4.8 indicate that the most effective choice of algorithmic depth $m = m_k$ may be to have it increase through the simulation based on the magnitude of the residual. We can think of the iteration as progressing through initial, pre-asymptotic, and asymptotic regimes, as we describe below. The different regimes can depend on the scaling of data and choice of initial iterates. Damping (not explicitly discussed here; see, for instance, [50]) may be necessary to see a reduction in the first-order residual terms in noncontractive settings, particularly in the initial regime. It is assumed here that the problem dimension n is significantly larger than the number of iterations allowed to solve the problem, and a "large" value of m_k is still small compared with n. The following 3-phase and 2-phase approaches are demonstrated in the numerical experiments that follow.

3-phase approach

This approach is appropriate for problems where the initial residual is large enough so that an accumulation of higher-order terms can easily prevent convergence. This method is demonstrated on the p-Laplacian in Section 4.5.

- Initial regime: The residual w_k and difference between iterates e_k may be large (in norm), for instance $\mathcal{O}(1)$ or greater. The depth m_k should be chosen small (for instance between 0 and 2), as the accumulation of higher-order terms on the right-hand side of (4.2.2) (cf. (4.3.7)) can prevent convergence.

- Preasympototic regime. The residual or difference between iterates is on the order of 10^{-1} or 10^{-2}. The depth m_k can safely be increased as $\|w_k\|$ decreases, either to convergence tolerance or until the asymptotic regime is reached.

- Asympototic regime. The residual is sufficiently small so that the higher-order terms of (4.2.2) are negligible, regardless of whether $\sqrt{1-\theta_k^2}$ is close to 1. The depth m_k can be increased but should be only to the point that it has an impact on decreasing the gain θ_k.

In the results shown below for the steady Navier-Stokes equations, simply choosing a large depth m (by which $m_k = k - 1$, either up to some maximum m or up to convergence) can be effective in the case that the initial residual is not greater than $\mathcal{O}(1)$ and drops sufficiently rapidly through the initial iterations. This is essentially the strategy above, starting in the pre-asymptotic, rather than the initial regime. In this example it is also demonstrated that switching to Newton iterations in the asymptotic regime can yield rapid convergence.

2-phase approach

This method is appropriate for problems with a moderately scaled initial residual, on the order of $\mathcal{O}(1)$, and is demonstrated in Section 4.5 on a nonlinear Helmholtz equation.

- Preasymptotic regime. The depth m_k is kept at a small to moderate value (2 to 5), until the residual drops below a given threshold, on the order of 10^{-2} or 10^{-3}.

- Asymptotic regime. The depth m_k is increased to a higher steady level, for instance $m_k = 10$. This allows smaller factors of the optimization gain θ_k due to a better solution of the least-squares problem at the point where the residual is small enough that the increased weight and accumulation of higher-order terms does not interfere with convergence.

Angle filtering and verification of (4.1.7) on sufficient linear independence of the columns of F_k

It makes sense both numerically and theoretically to solve the least-squares problem at each stage by means of an economy QR decomposition. Given the large number of degrees of freedom n in comparison to the algorithmic depth m, forming the decomposition and solving the least-squares system has a negligible effect on total solution time; in each of our examples given below, the total runtime is dominated by the linear system solves and/or Jacobian assembly. Angle filtering is summarized in the following algorithm, which would be performed between lines 5 and 6 of Algorithm 1.1. This algorithm is restated in Chapter 5 as Algorithm 5.2 as part of a complete filtering strategy that ensures the matrix F_k is well conditioned at each iteration.

ALGORITHM 4.1. Angle filtering (version 1).
Given a threshold $c_s \in (0,1)$

1: Compute the economy QR decomposition $F_k = \hat{Q}\hat{R}$
2: Compute $\sigma_i = |r_{ii}|/\|f_i\|$, $i = 2, \ldots, m_k$, where f_i is column i of F_k, and r_{ii} is the corresponding diagonal entry of $\hat{R}$
3: Remove any columns i of E_k and F_k for which $\sigma_i < c_s$
4: Update m_k with the new number of columns of F_k
5: **if** any columns were removed **then**
6: Recompute $F_k = \hat{Q}\hat{R}$
7: **end if**

For any threshold value in $c_s \in (0,1)$, Assumption 4.3 is necessarily satisfied, and the algorithm ensures the most recent (first) column of F_k is always used since $r_{11}/\|f_1\| = 1$. Algorithm 4.1 is demonstrated in section 4.5 on a finite element discretization of the p-Laplace equation, with p close to one.

4.5 ▪ Numerical experiments

The test problems in this section will illustrate the theory and the practical guidance given above. We demonstrate angle filtering and dynamic depth selection strategies, and we demonstrate the practical efficiency of AA for the p-Laplace equation (see Section 3.6); a nonlinear Helmholtz equation (see Section 3.2); the steady Navier-Stokes equations in two and three dimensions (see Section 3.3); the Boussinesq equations (see Section 3.8); and, a regularized model for non-Newtonian Bingham flow (see Section 3.7).

4.5.1 ▪ p-Laplace equation

We first consider a finite element discretization of the regularized p-Laplace equation (3.6.1) subject to homogeneous Dirichlet boundary conditions and $\epsilon = 10^{-14}$, $p = 1.06$ (cf. [47]), and $c = \pi$. We compare results using two uniform triangulations of $[0,2] \times [0,2]$: a coarser mesh with 128 subdivisions in the x and y directions which has 66,049 total degrees of freedom and a finer mesh with 256 subdivisions in each direction which has 263,169 degrees of freedom. We use P_2 elements and the Picard iteration (3.6.2). The simulations are run using a Python implementation of the FEniCS finite element library [4]. Each simulation was started from initial iterate $u_0 = xy(x-1)(y-1)(x-2)(y-2)$ (cf. [29]) and run to a residual tolerance of $\|w_k\| < 10^{-10}$, where an l_2 norm is used for both the convergence tolerance and the optimization.

With the given parameters, the defined fixed-point operator is not globally contractive, but the Picard iteration does converge essentially linearly as it approaches the solution. On each discretization we compare applying AA with fixed depths $m = 0, 2, 4, 8$, with and without angle-filtering, and dynamically updating the depth m_k by defining

$$\tilde{m}_k := \text{ceil}(-\log_{10} \|w_{k+1}\|) \quad \text{and } m_k = \psi_{n,N}(\tilde{m}_k) := \begin{cases} n, & \tilde{m}_k \le n, \\ \tilde{m}_k, & n < \tilde{m}_k < N, \\ N, & \tilde{m}_k \ge N. \end{cases} \tag{4.5.1}$$

Setting the depth by (4.5.1) implements the 3-phase approach described in Subsection 4.4.

Figure 4.1 shows residual histories for the coarser mesh (top row) and finer mesh (bottom row) using constant depths $m = \{0, 2, 4, 6\}$ (left plots) and $m = \{0, 8\}$ together with dynamically chosen depths $\psi_{0.0,8}$ and $\psi_{1,8}$ given by (4.5.1), each with and without angle filtering (AF), with a threshold of $c_s = 0.45$.

Comparing the top and bottom plots on both sets, we see that for greater constant algorithmic depths $m = 6, 8$, the iteration stays trapped longer in the preasymptotic regime on the finer mesh. The AF approach allows the iteration to quickly reduce the residual and attain asymptotic convergence. In the right two plots, we see the dynamic depth strategy with and without filtering works to similar effect. These results support the analysis shown in this chapter: when the residual is too large, a large depth allows the higher-order terms to collectively prevent convergence.

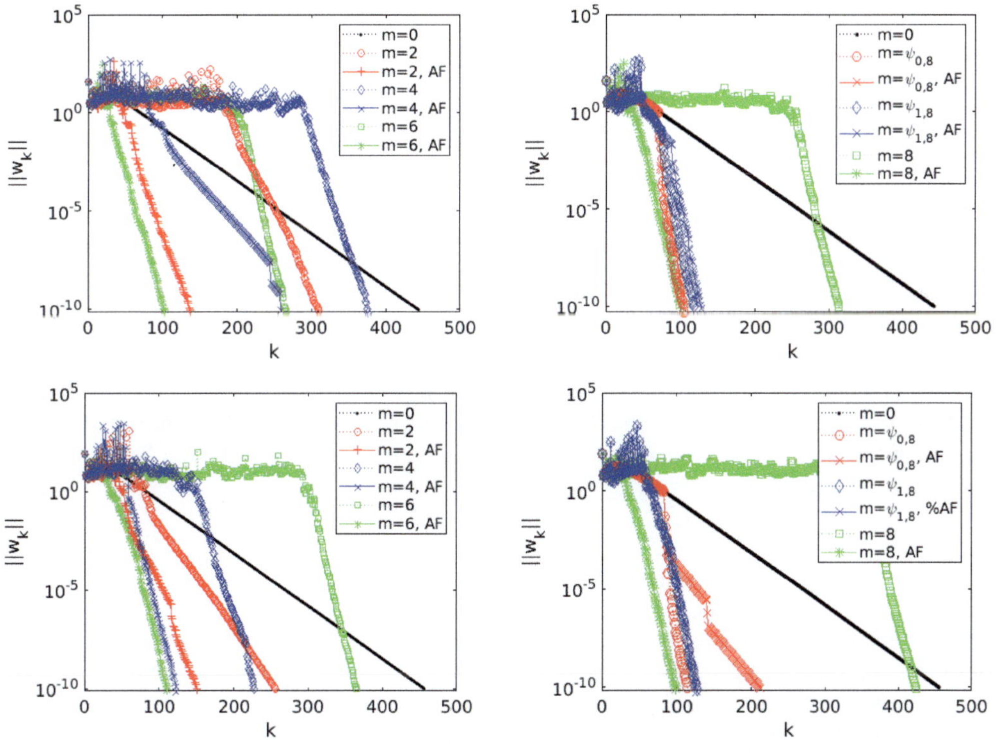

Figure 4.1. *Residual history for AA applied to* (3.6.2) *with and without angle-filtering (AF), with threshold $c_s = 0.45$. The top two plots use the 128×128 subdivision mesh, and the bottom two plots use the 256×256 subdivision mesh. The left two plots use constant algorithmic depths $m = \{0, 2, 4, 6\}$. The right two plots use the constant depths $m = \{0, 8\}$ and the dynamically selected depths $\psi_{0,8}$ and $\psi_{1,8}$ as given by* (4.5.1).

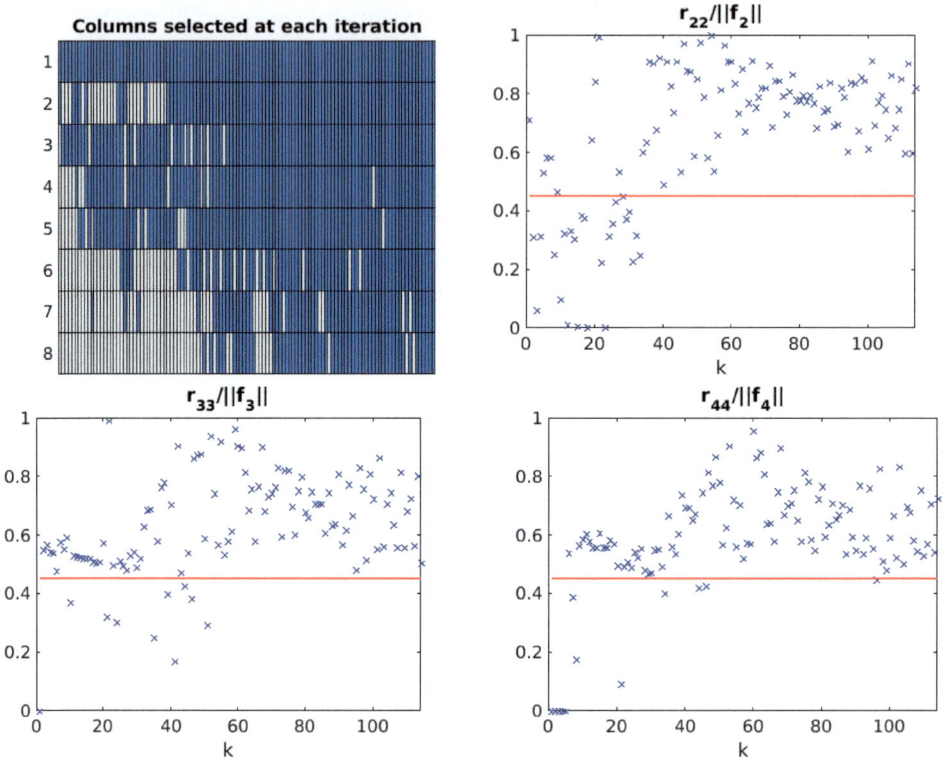

Figure 4.2. *Top left: heatmap of the columns chosen at each iteration for the constant $m = 8$ simulation on the 128×128 mesh using AF. The remaining three plots show the critical quantity $r_{ii}/\left\|f_i\right\|$, for $i = 2, 3, 4$, for the same simulation.*

It is interesting to notice that the best convergence or close to it can be achieved by more than one method.

To visualize the choice of the AF threshold, Figure 4.2 shows a heatmap of the columns chosen at each iteration for the constant $m = 8$ simulation on the 128×128 mesh, using AF. The remaining three plots show the critical quantity $r_{ii}/\left\|f_i\right\|$, for $i = 2, 3, 4$, for the same simulation. This shows how the second column has several "near-linear dependence" situations early in the preasymptotic regime. In these visualizations, everything that falls below the red line represents a column that gets filtered out.

4.5.2 ▪ Nonlinear Helmholtz equation

Next, we consider the nonlinear Helmholtz (NLH) equation, Picard iteration, and discretization as described in Section 3.2. This test uses $\epsilon = 0.22$, for which the Picard iteration does not converge. Figure 4.3 shows the fixed-point iteration ($m = 0$) with varying levels of relaxation (damping); this illustrates that (uniform) relaxation alone is not sufficient for convergence. In Figure 4.4, results of AA applied to the iteration using relaxation parameter $\beta_k = \beta = 0.3$ are shown for $m = 1, 3, 5, 10$, all of which converge. The plot of k *vs.* θ_k shows a clear reduction in gain factors θ_k as the depth m increases. Comparing convergence histories for varying depths m, none of the depths tested show monotonic decrease, particularly in the preasymptotic regime. Depth $m = 10$, which becomes nearly monotone in the asymptotic regime, has gain values

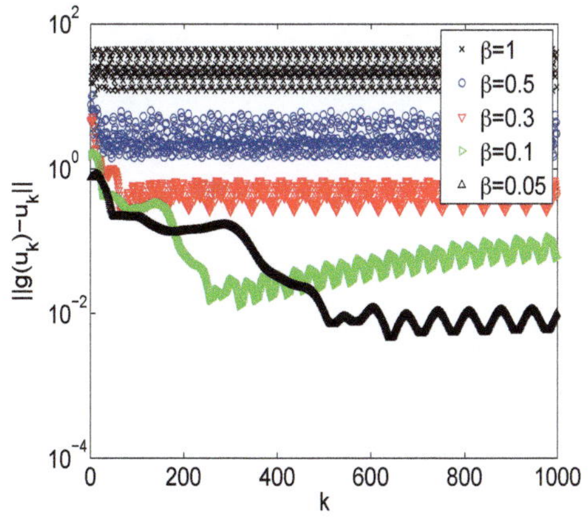

Figure 4.3. *Results of the NLH test with $\epsilon = 0.22$ and $m = 0$ demonstrating that decreasing a fixed damping factor β does not induce convergence of the fixed-point iteration. [88], distributed under CC BY 4.0 DEED License (https://creativecommons.org/licenses/by/4.0/).*

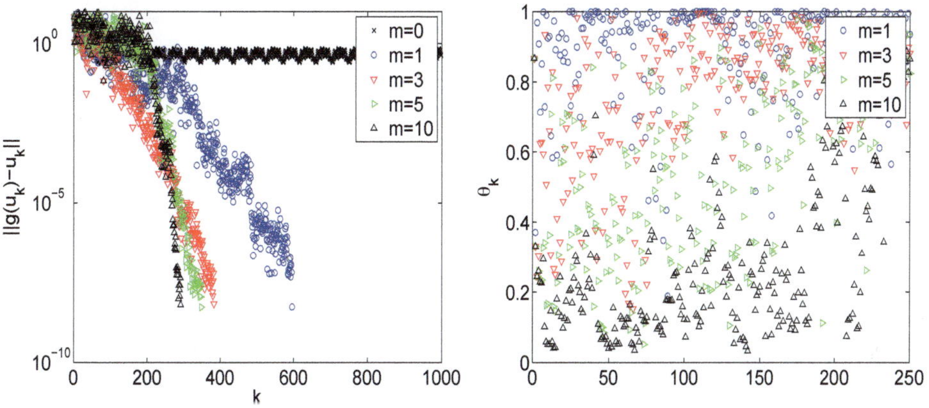

Figure 4.4. *Results of the NLH test with $\epsilon = 0.22$ as convergence of the nonlinear residual (left) for $\beta_k = \beta = 0.3$ and varying m, and θ_k for varying m (right). [88], distributed under CC BY 4.0 DEED License (https://creativecommons.org/licenses/by/4.0/).*

generally less than 0.6, whereas depth $m = 1$ which is far from monotone has gain values that return to nearly 1 throughout the first 250 iterations shown in Figure 4.4 on the right.

The next results, shown in Figure 4.5, use a heuristic strategy for updating m. This strategy is based on the observation that depth $m = 3$ gives a faster initial decrease in the residual, and $m = 10$ gives the fastest eventual decrease. Here, depth m_k is switched from $m_k = 3$ to $m_k = 10$ on the condition of a sufficiently small residual, where the tolerance is set at 0.005. The depth-switching approach yields substantially faster convergence than either constant-depth strategy. This is again consistent with the theory, as larger higher-order terms play a greater role earlier in the iteration history and moreso at greater algorithmic depths. Once the higher-order terms are sufficiently small (attained through a sequence of sufficiently small gain values), the decrease in gain θ_k for greater depths m yields better performance as the residual is small enough to be dominated by the first-order term even as the number and weight of the higher-order terms increase.

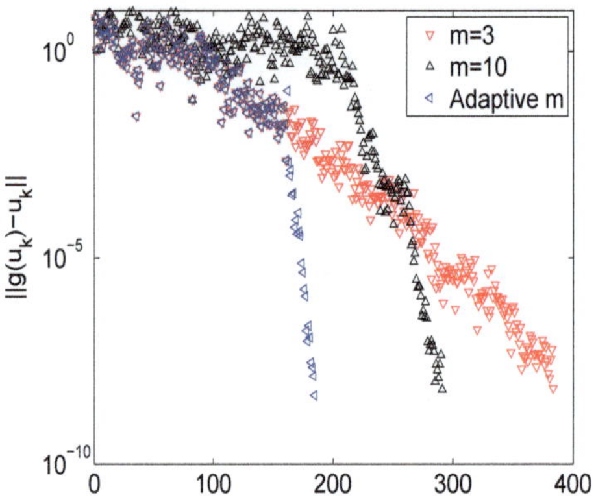

Figure 4.5. *Results of the NLH test with* $\epsilon = 0.22$ *as convergence of the nonlinear residual (left) for* $\beta_k = \beta = 0.3$ *and* $m = 3$, $m = 10$, *and an heuristic strategy where* $m = 3$ *at first but switches to* $m = 10$ *once the nonlinear residual is sufficiently small. [88], distributed under CC BY 4.0 DEED License (https://creativecommons.org/licenses/by/4.0/).*

4.5.3 ▪ Navier-Stokes equations

2D Navier-Stokes driven cavity

We test the Navier-Stokes equations for the 2D driven cavity problem with $Re = 5000$. We follow the setup, discretization, and Picard iteration from Section 3.3 and use Taylor-Hood elements on a $h = 1/64$ barycenter refined uniform mesh, no relaxation, and varying m; we test with no AA ($m = 0$), and m =1,2,5 and 10. Convergence plots are shown in Figure 4.6, and we observe no convergence without AA, slow convergence ($\sim$175 iterations) for $m = 1$, significantly faster convergence (50 iterations) for $m = 2$, and slightly faster still by increasing m to 5 and then $m = 10$.

Also shown in Figure 4.6 is a plot for $m = 1$ of residuals $\|w_k\|_1 = \|\nabla w_k\|$ and optimization gains θ_k (zoomed in to just iterations 10 to 30) versus iteration number. Here we observe the relationship between the residual decrease (or increase) and the optimization gain factor predicted

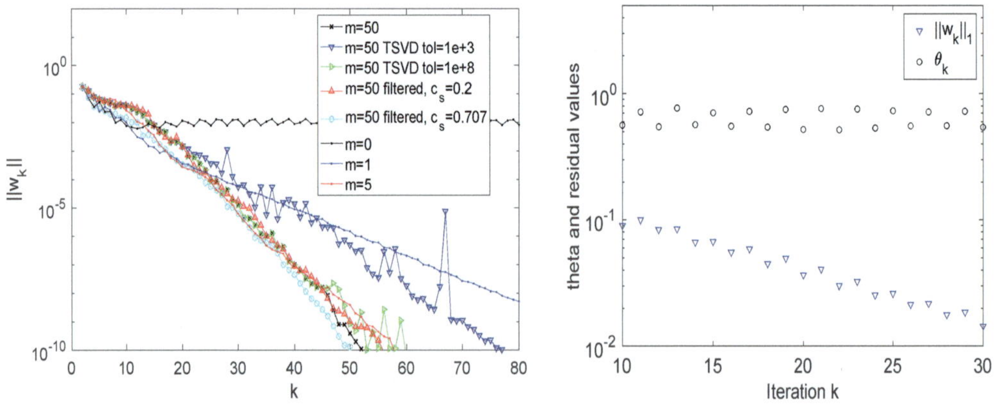

Figure 4.6. *Shown above are results for the 2D driven cavity. At left is a plot of nonlinear residuals versus iteration number for varying m, and at right are* θ_k *gain factors and nonlinear residuals* $\|w_k\|_1$ *versus iteration numbers 10 to 30 for* $m = 1$.

by Theorems 4.4 and 4.5 that the linear convergence rate is scaled by the optimization gain factor. In particular, for the larger values of θ_k, the kth residual increases, but for the smaller values of θ_k the kth residual decreases.

3D Navier-Stokes driven cavity

We also test the 3D driven cavity benchmark test problem for the steady Navier-Stokes equations (NSE), as described in Section 3.3 and with discretization and Picard iteration described therein. Recently, higher Re have been considered, but as a *time dependent flow*, in an attempt to find the first Hopf bifurcation (around $Re \approx 2000$) where the flow becomes oscillatory and will not converge to a steady state [33, 55]. Solvers for $Re \geq 1000$ typically do not directly solve this system; instead, approaches are to use continuation methods in the Re or time dependent methods and drive them to a steady state. While such techniques can be somewhat effective, they are typically quite slow. We show herein that with AA, the Picard method is able to converge directly for Re as high as 3100.

For this test problem, as shown in [91], with an initial guess of $u_0 = 0$, the Picard method does not converge when $Re \geq 400$. Hence, for $Re \geq 400$, Picard iterations for steady solutions are not globally contractive. In fact, applying AA to Picard with $m = 1$ fails as well, although convergence is attained with depths $m = 2, 3, 4$, as demonstrated in [91]. To show the effectiveness of AAPicard, considerably higher Reynolds numbers are considered here: results are presented for $Re = 2500$ and $Re = 3100$, far beyond the range where the Picard iteration is contractive and, moreover, well past the first Hopf bifurcation [33, 55] . Thus the method is converging to steady solutions in a time dependent regime, which from a mathematical point of view is interesting in itself. As discussed in [3], such solutions can serve as base-flow solutions in instability studies and flow control.

The $Re = 2500$ tests show different choices of the depth m, including the largest possible ($m_k = k - 1$), with no relaxation ($\beta_k = \beta = 1$). Results are shown in Figure 4.7. For $m \leq 50$, convergence is not achieved after 500 iterations. For $m = 100, 150$, and $m = k - 1$, the method does converge. It appears that the stability of the NSE Picard iteration [64] bounds the magnitude of any residual, and the improved analysis herein shows that higher-order terms are all scaled by the latest residual, which together allows the method to benefit from the small gain factor θ_k that comes from a greater algorithmic depth m ($m \geq 100$ creates gain factors θ_k that get to 0.25 and below). Notably, choosing m as large as possible, $m_k = k - 1$ gives the fastest convergence.

Finally, a combination of AAPicard with Newton was tested (the Newton iteration differs from the Picard in that the term $(u_{k+1} - u_k) \cdot \nabla u_k$ is added to the left side of the iteration from Section 3.3 [74]). Additionally, a line search was used in the Newton iterations. The results shown used $m_k = k - 1$ for the initial AAPicard iterations and switched to Newton once the nonlinear residual reached a sufficiently low tolerance. For an H_0^1-norm tolerance of 1, the method failed to converge, but for tolerances of 0.1 and 0.01, the method converged, and much faster than AAPicard does on its own (see the top plot in Figure 4.7).

With this technique, the solver attained convergence up to $Re = 3100$ (using AAPicard with $m_k = k - 1$ and $\beta_k = \beta = 0.5$, up to a residual tolerance of 0.03, and then switching to Newton with a line search). With this method, 217 iterations were needed to converge to a tolerance of 10^{-8} in the H_0^1-norm. With a continuation method that improves the initial guess, solutions at even higher Re can be obtained.

4.5.4 ▪ Boussinesq equations

We now test AA applied to the discretized Picard iteration for the steady Boussinesq equations, following the problem and discretization setup from Section 3.8. The natural norm for measuring

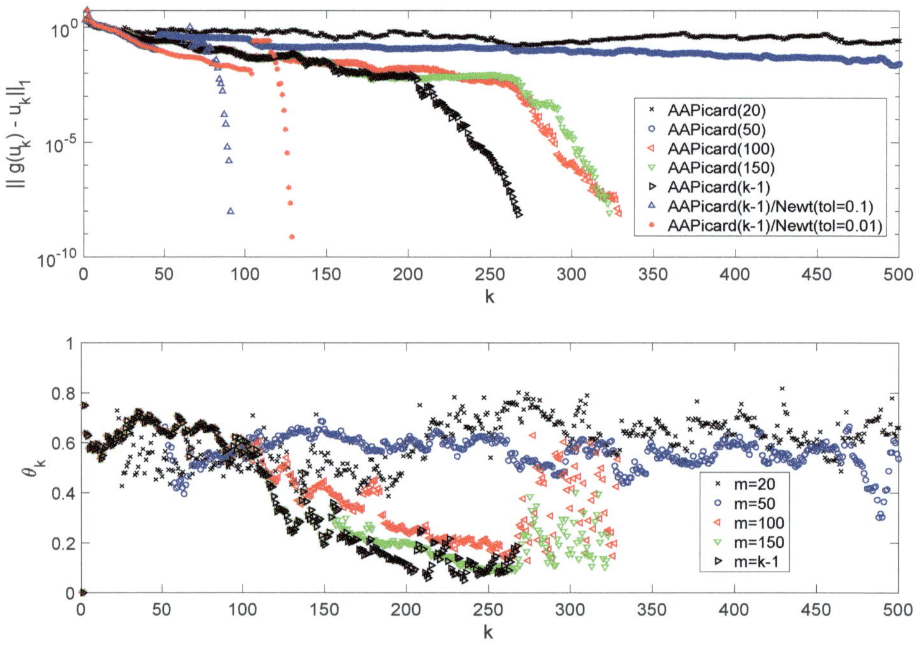

Figure 4.7. *Top: convergence of AAPicard with varying m with and without a switch to Newton. Bottom: gain factors θ_k for varying m. [88], distributed under CC BY 4.0 DEED License (https://creativecommons.org/licenses/by/4.0/).*

convergence for this problem is the B-norm, which is defined by $\|(\cdot,\cdot)\|_B : (V_h, W_h) \to \mathbb{R}$:

$$\|(v, w)\|_B := \sqrt{\nu\|\nabla v\|^2 + \kappa\|\nabla w\|^2}.$$

Heated cavity problem

We first consider the differentially heated cavity problem, with varying Rayleigh numbers. Our discretization uses (P_2, P_1^{disc}) velocity-pressure Scott-Vogelius elements and P_2 temperature elements. The mesh is created by starting with a uniform triangulation, refining along the boundary, and then applying a barycenter refinement. This yielded a discretization with 89,554 total dof. We show results below for varying Rayleigh numbers, which come from using parameters $\nu = 0.01$ and $\kappa = 0.01$ and varying Ri.

In our tests, we consider both constant algorithmic depths m, and also a 2-stage strategy that uses a smaller m (1 or 2) when the nonlinear residual is larger than 10^{-3} in the B-norm, and $m = 20$ when the residual is smaller. The 2-stage approach is motivated by Theorem 4.5 (and Theorem 6.11, which is specific to the Picard iteration for the Boussinesq system), which suggests that greater depths m can be detrimental early in the iteration due to the accumulation of non-negligible higher-order terms in the residual expansion. Once the residual is sufficiently small, then the reduction of the first-order terms from a greater depth m can be enjoyed without noticeable pollution from the higher-order terms, which essentially results in an improved linear convergence rate.

We test here the AA-Picard iteration for Boussinesq, for varying m and β, and for several Rayleigh numbers: $Ra = 10^5,\ 5\cdot 10^5,\ 10^6,\ 2\cdot 10^6$. For each Ra, we tested with $m = 0$ and each fixed damping parameter $\beta = 0.05, 0.1, 0.15, \ldots, 1$ and then used the best β from all tests with that Ra. Respectively, for $Ra = 10^5,\ 5\cdot 10^5,\ 10^6,\ 2\cdot 10^6$, these parameters were $\beta = 0.3,\ 0.05,\ 0.05,\ 0.05$.

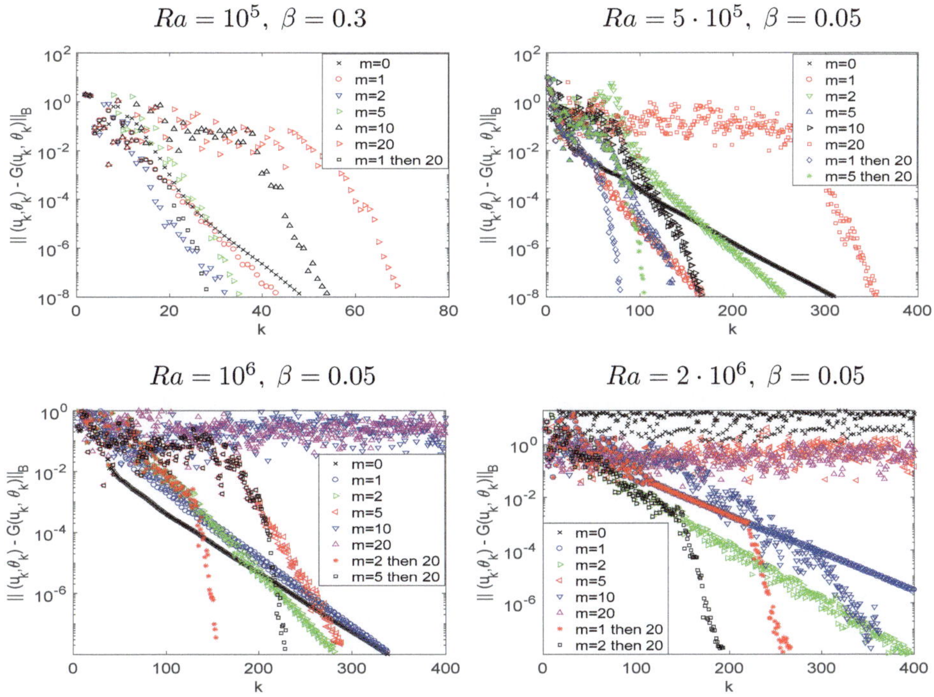

Figure 4.8. *Convergence results for the differentially heated cavity with varying m for several Rayleigh numbers. [92], distributed under CC BY 4.0 DEED License (https://creativecommons.org/licenses/by/4.0/).*

Convergence results are shown in Figure 4.8, displayed in terms of the B-norm of the nonlinear residual. The usual Picard iteration, i.e., $m = 0$ and $\beta = 1$, did not converge for any of these Ra numbers; after 500 iterations, the B-norm was still larger than 10^{-1}. With the appropriately chosen relaxation parameters, the Picard iteration did converge for each case except when $Ra = 2 \cdot 10^6$. Results for $Ra = 10^5$ are shown in the top left of Figure 4.8 for varying m and fixed $\beta = 0.3$. Here, we observe that the unaccelerated method converges rather quickly, and the accelerated methods that converged faster were $m = 1$ and $m = 2$ and the 2-stage method that uses $m = 1$ and then switches to $m = 20$, with the 2-stage method giving the best results. AA-Picard with $m = 5,\ 10,\ 20$ all converged but slower than if no acceleration was used. Results from higher Ra tell a similar story: the best results come from using a 2-stage choice of m, using a lesser depth at the beginning of the iteration and a greater depth once the residual is sufficiently small.

We also include convergence results in Table 4.1 for no relaxation ($\beta = 1$), $Ra = 10^4$ and 10^5, $m = 1$ and $m = 5$ for varying meshes $h = \frac{1}{16}, \frac{1}{32}, \frac{1}{64}$. The purpose of this is to

Table 4.1. *Number of Anderson accelerated Picard iterations required for convergence for the Boussinesq heated cavity problem. [92], distributed under CC BY 4.0 DEED License (https://creativecommons.org/licenses/by/4.0/).*

$\downarrow Ra\ h \rightarrow$	$\frac{1}{16}$	$\frac{1}{32}$	$\frac{1}{64}$	$\frac{1}{16}$	$\frac{1}{32}$	$\frac{1}{64}$
	$m = 1$			$m = 5$		
10^4	19	18	18	13	13	12
10^5	76	81	79	42	38	43

investigate the effect of the mesh width on the convergence, and we observe nearly perfect mesh independence.

Boussinesq on a complex domain

We now present results from a second test that has a more complex domain, as described in Section 3.8. We use a barycenter refined uniform triangulation mesh, together with (P_2, P_1^{disc}) Scott-Vogelius velocity-pressure elements, and P_2 for temperature which provides 70,068 total dof. We use parameters $\nu = 0.01$, $\kappa = 0.01$ and then choose the Richardson number Ri to create Rayleigh numbers $Ra = 10^4$ and 5×10^4; plots of resolved solutions are shown in Figure 3.7.

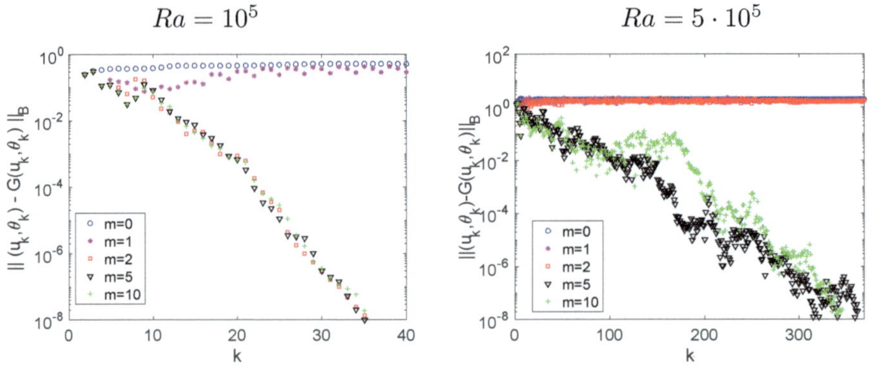

Figure 4.9. *Convergence plots of Anderson accelerated Picard iteration for Boussinesq on a complex domain, with Rayleigh numbers $Ra = 10^4$ (left) and 5×10^4 (right) and AA depth $m = 0, 1, 2, 5, 10$. [92], distributed under CC BY 4.0 DEED License (https://creativecommons.org/licenses/by/4.0/).*

For each Ra, we run the Anderson accelerated Picard iteration using varying m and $\beta = 1$. Plots for convergence are shown in Figure 4.9, and in both cases we observe that the unaccelerated Picard iteration ($m = 0$) fails to converge, as does $m = 1$. For $Ra = 10^4$, tests with $m \geq 2$ all converged, and $m = 2, 5, 10$ converged at about the same rate. For $Ra = 5 \times 10^4$, $m = 2$ fails but $m = 5$ and $m = 10$ both converge. As in the previous test, AA enables convergence of the Picard iteration if it is applied with a sufficiently large algorithmic depth m. However, if the depth is too great, pollution from earlier iterates interferes with convergence, at least until the error is sufficiently small. This can be seen in Figure 4.9 by the reduced regularity in the earlier stages of convergence with $m = 10$, as compared with $m = 5$.

4.5.5 ▪ Bingham equations

This section presents the results of three numerical tests that illustrate the theory above for AA applied to the discretized Picard iteration for regularized Bingham equations (3.7.3). In all tests, AA provides significantly faster convergence than without it, especially with small ϵ.

Analytical test

The flow between two parallel plates is one of the known analytical test cases for Bingham problems. In two dimensions, the analytical solutions of Stokes-type Bingham equations are given by

$$u_1 = \begin{cases} \frac{1}{8}\left[(1-2\tau_s)^2 - (1-2\tau_s-2y)^2\right], & 0 \le y < \frac{1}{2} - \tau_s, \\ \frac{1}{8}(1-2\tau_s)^2, & \frac{1}{2} - \tau_s \le y \le \frac{1}{2} + \tau_s \\ \frac{1}{8}\left[(1-2\tau_s)^2 - (2y-2\tau_s-1)^2\right], & \frac{1}{2} + \tau_s < y \le 1. \end{cases}, \quad u_2 = 0, \text{ and } p = 0, \tag{4.5.2}$$

The rigid (or plug) region $\{y \in \Omega \mid \frac{1}{2} - \tau_s \le y \le \frac{1}{2} + \tau_s\}$ is the kernel moving at constant velocity. We choose $\tau_s = 0.3$. The discretization uses (P_2, P_1) Taylor-Hood elements on a uniform triangular mesh. We take $\mu = 1$ and external force $f = 0$ and perform Anderson accelerated Picard iterations with depth $m = 0$ (no acceleration), $1, 2, 5$, and 10.

We display the number of iterations that reduce the relative L^2 residual of the velocity by 10^{-8} for varying depths m, mesh sizes h, and regularization parameters ϵ in Table 4.2. When $\epsilon \to 0$ and/or the mesh width decreases, the required number of iterations increases (this increase is expected from the analysis in section 6.3). Also, AA provides better convergence results as we increase the depth. This improvement is more apparent in smaller values of ϵ, which is required to obtain an accurate solution. In the case of $m = 0$ (without AA), the numbers of iterations are substantially higher; however, with AA they decrease significantly. The fastest convergence is obtained with depths $m = 5$ and 10, as seen in Table 4.2.

Table 4.2. *Number of Anderson accelerated Picard iterations required for reducing the residual by 10^{-8} with different depths for the regularized Bingham equation when $\tau_s = 0.3$ and h and ϵ are changing in the analytical test case. Reprinted from [90], with permission from Springer Nature.*

$\downarrow h\ \varepsilon \rightarrow$	10^{-1}	10^{-2}	10^{-3}	10^{-4}	10^{-5}		10^{-1}	10^{-2}	10^{-3}	10^{-4}	10^{-5}
$m = 0$						$m = 1$					
1/8	10	34	42	88	101		8	22	24	31	43
1/16	10	34	80	171	274		9	21	36	60	94
1/32	11	27	98	258	296		9	21	45	95	114
1/64	11	35	108	184	184		9	21	46	57	49
1/128	11	35	106	306	408		9	21	43	71	69
$m = 5$						$m = 10$					
1/8	7	14	16	22	26		7	14	16	21	25
1/16	7	15	24	30	33		7	15	23	28	33
1/32	8	16	29	36	41		8	15	25	32	34
1/64	8	16	29	34	40		8	15	26	34	39
1/128	8	16	29	48	50		8	15	26	39	51

2D driven cavity

We next test the Anderson accelerated Picard iteration for the regularized Bingham equations on a lid-driven cavity problem. The problem setup is the same as for 2D Navier-Stokes driven cavity above. The discretization uses (P_2, P_1) Taylor-Hood elements on a uniform mesh.

Figure 4.10 shows the number of iterations of the Anderson accelerated Picard iteration with varying depth m and regularization parameter ϵ when $h = 1/64$ and yields stresses $\tau_s = 2$ and $\tau_s = 5$. Iterations were run until the relative L^2 velocity residual fell below 10^{-8}. As ϵ becomes smaller, the required number of iterations increases. As illustrated in Figure 4.10, the original (unaccelerated) Picard method converges very slowly compared to the accelerated method. However, as the value of depth m is increased up to 10, the accelerated method converges much faster.

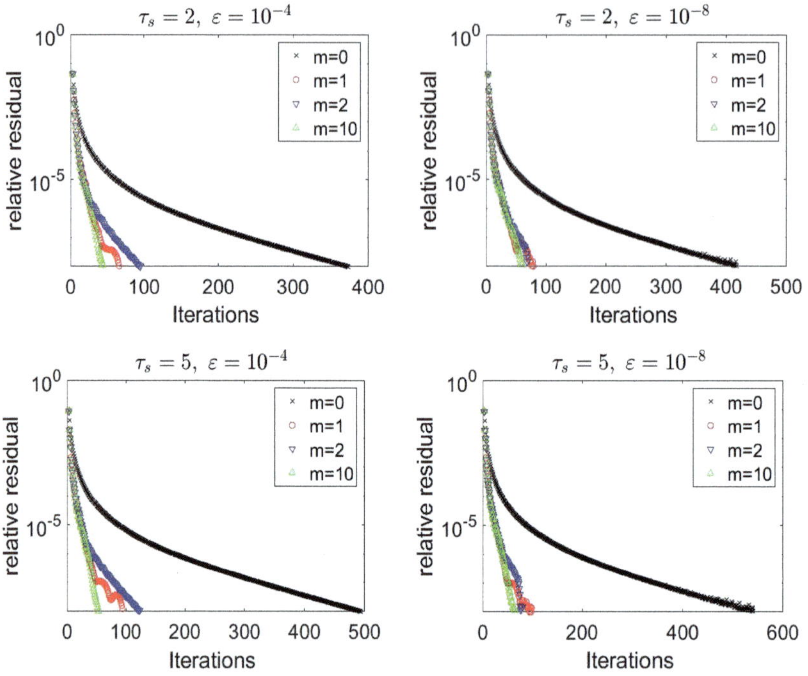

Figure 4.10. *Convergence for different* $\varepsilon = 10^{-4}, 10^{-8}$*(left to right) when* $\tau_s = 2$ *(top) and* $\tau_s = 5$ *(bottom) with varying* m*. Reprinted from [90], with permission from Springer Nature.*

Numerical verification of Assumption 4.2

The application of the AA theory above to the Picard iteration for the regularized Bingham equations relies on the satisfaction of Assumptions 4.1 and 4.2. We analytically verify Assumption 4.1 in Section 6.3. Assumption 4.2 is satisfied if the Jacobian of the fixed-point iteration (which we denote now by G) does not degenerate. To demonstrate the satisfaction of this assumption, here we calculate the ratio $\frac{\|w_k - w_{k-1}\|}{\|u_k - u_{k-1}\|}$ for two numerical tests using constant m to show the σ_k (the minimum of the k through $k - m_k$ ratios) is bounded well above 0. Results for the analytical test problem and 2D driven cavity are shown in Figure 4.11. We observe that the ratios never get close to 0 and in general get larger for larger m.

Proof of technical Lemma 4.13

We conclude this chapter with the proof of the technical Lemma 4.13, bounding the entries of $\hat{R}^{-1}$, where $\hat{R}$ is the upper triangle matrix in an economy QR factorization of matrix A.

Proof. The proof follows by induction on the submatrix formed by the first p rows and columns of $\hat{R}$ and then by induction indexing up the entries of the right-most column. Let $R_p = R(1 : p, 1 : p)$, the upper-left $p \times p$ block of $\hat{R}$, with inverse S_p.

The off-diagonal entries r_{ij} of $\hat{R}$ are given by $r_{ij} = (q_i, a_j) = \|a_j\| \cos(q_i, a_j)$, and by Proposition 4.12 the diagonal entries are given by $r_{ii} = \|a_i\| \, |\sin(a_i, \mathcal{A}_{i-1})|$, following the convention that the columns of $\hat{Q}$ are chosen so the r_{ii} are positive.

For the trivial case of $p = 1$, $R_1 = r_{11}$, and $s_{11} = 1/r_{11} = 1/\|a_1\|$. By Proposition 4.11, to compute the inverse of R_2 it remains to compute s_{22} and s_{12}. It is useful here to state the

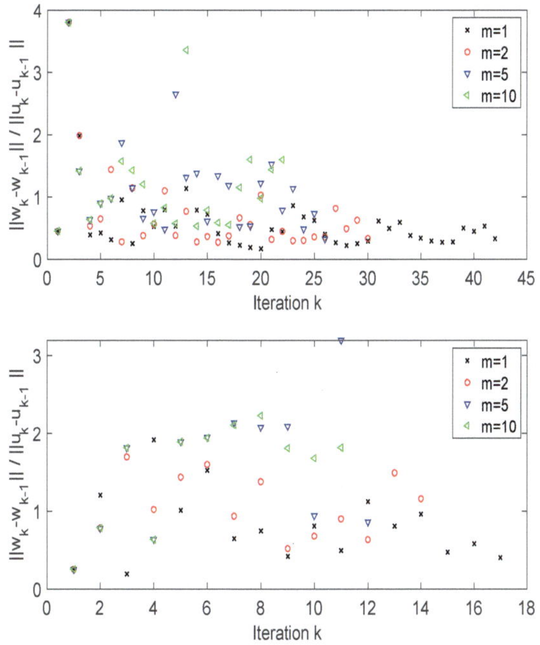

Figure 4.11. *Shown above are ratios of* $\frac{\|w_k - w_{k-1}\|}{\|u_k - u_{k-1}\|}$ *for (top) the analytical test using* $h = 1/32$, $\tau_s = 0.3$, $\varepsilon = 10^{-3}$ *and (bottom) for 2d lid driven cavity* $h = 1/64$, $\tau_s = 2$, $\varepsilon = 10^{-1}$, *both with varying* m. *Reprinted from [90], with permission from Springer Nature.*

inversion formula for entries of the right-most column (index p) as

$$s_{pp} = \frac{1}{r_{pp}},$$

$$s_{kp} = -\frac{1}{r_{kk}} \sum_{j=1}^{p-k} r_{k,k+j} s_{k+j,p} = -\frac{1}{r_{kk}} \sum_{j=1}^{p-k} \|a_{k+j}\| \cos(q_k, a_{k+j}) s_{k+j,p}, \quad k < p. \tag{4.5.3}$$

For $p = 2$, the inversion formula (4.5.3) and expression (4.3.19) for the diagonal entries yield $s_{22} = 1/r_{22} = 1/(\|a_2\| \, |\sin(a_2, q_1)|)$. Then, by the hypotheses of the lemma, $s_{22} \le 1/(\|a_2\| \, c_s)$. Using (4.5.3), the off-diagonal entry then satisfies $s_{12} = -\|a_2\| \cos(q_1, a_2) s_{22}/r_{11}$, yielding $|s_{12}| \le c_t/(\|a_1\| \, c_s)$. Hence for $p = 2$ the result holds. Continue by induction on p, assuming the result holds for $q = 1, \dots, p-1$. Then, for $q = p$,

$$R_p = \begin{pmatrix} R_{p-1} & r_{1p} \\ & \vdots \\ 0 & r_{pp} \end{pmatrix}.$$

By (4.5.3), Proposition 4.12, and the hypotheses of the lemma, $\|s_{pp}\| \le 1/(\|a_p\| \, c_s)$.

Similarly, by (4.5.3), $\|s_{p-1,p}\| \le c_t/(\|a_{p-1}\| \, c_s^2)$. This satisfies the base step on the inner induction, up to row p of S_p. Assuming the bound of (4.3.21) for s_{ip} holds for $i = p-1$ down to $i = k+1$, it suffices to show the result for $i = k$. By (4.5.3) and the inductive hypothesis,

$$|s_{kp}| = \left| \frac{1}{r_{kk}} \sum_{j=1}^{p-k} \|a_{k+j}\| \cos(q_k, a_{k+j}) s_{k+j,p} \right| \le \frac{1}{r_{kk}} \left(\sum_{j=1}^{p-k-1} \frac{c_t^2 (c_t + c_s)^{p-(k+j)-1}}{c_s^{p-(k+j)+1}} + \frac{c_t}{c_s} \right).$$

Setting $n = p - k$,

$$|s_{kp}| \leq \frac{1}{r_{kk}} \left(\sum_{j=1}^{n-1} \frac{c_t^2 (c_t + c_s)^{n-j-1}}{c_s^{n-j+1}} + \frac{c_t}{c_s} \right) = \frac{c_t}{c_s^n r_{kk}} \left(\sum_{j=1}^{n-1} c_t (1 + c_s)^{n-j-1} c_s^{j-1} + c_s^{n-1} \right). \tag{4.5.4}$$

Rearranging the terms in the sum shows

$$\begin{aligned}
&\sum_{j=1}^{n-1} c_t (c_t + c_s)^{n-j-1} c_s^{j-1} + c_s^{n-1} \\
&\quad = \sum_{j=1}^{n-2} c_t (c_t + c_s)^{n-j-1} c_s^{j-1} + (c_t + c_s) c_s^{n-2} \\
&\quad = \sum_{j=1}^{n-3} c_t (c_t + c_s)^{n-j-1} c_s^{j-1} + (c_t + c_s)^2 c_s^{n-3} \\
&\quad \vdots \\
&\quad = c_t (c_t + c_s)^{n-2} + c_t (c_t + c_s)^{n-3} c_s + (c_t + c_s)^{n-3} c_s^2 \\
&\quad = c_t (c_t + c_s)^{n-2} + (c_t + c_s)^{n-2} c_s \\
&\quad = (c_t + c_s)^{n-1}.
\end{aligned} \tag{4.5.5}$$

Applying (4.5.5) and (4.3.19) to (4.5.4) allows

$$|s_{1p}| \leq \frac{c_t (c_t + c_s)^{p-2}}{\|a_1\| \, c_s^{p-1}} \quad \text{and} \quad |s_{kp}| \leq \frac{c_t (c_t + c_s)^{p-k-1}}{\|a_k\| \, c_s^{p-k+1}}, \quad k = 2, \ldots, p-1,$$

which completes the inductive step on k, and hence on p, and establishes the result. □

Chapter 5

Filtering for Anderson acceleration

The older you get, the better you get, unless you're a banana.
—Rose Nylund

In this chapter we address the issue of controlling the condition of the matrix F_k used in the AA least-squares problem of Algorithm 1.1, in order to improve the numerical stability of the algorithm. We will present a column filtering strategy from our work in [89] to efficiently control the condition of F_k that will not harm the AA convergence; and, in many cases it will noticeably improve convergence. Most of this chapter follows similarly to our paper [89].

In [53], both truncated singular value decomposition (TSVD) and Householder QR with column pivoting (rank revealing QR) are discussed as methods to address poorly conditioned least-squares problems that may arise in AA. These standard approaches to ill-conditioned linear least-squares problems will help address the issue of F_k conditioning; however, they may ultimately interfere with the convergence of AA as the columns of the iteration-k least-squares matrix F_k have a natural ordering based on the algorithmic age of the information they represent. Another approach to this problem is to add a Tikhonov-type regularization term to the least-squares problem. This approach is used, for instance, in [107] in the context of unconstrained optimization and in [109] in the context of deep reinforcement learning. While this regularization can be used to control the size of the optimization coefficients, the coefficients it selects are suboptimal and can neutralize the convergence rate advantages found from using AA to begin with [116].

To demonstrate the column filtering strategy in the numerical tests in this chapter, we will compare its performance with the standard TSVD approach in Section 5.3. The filtering algorithm often shows better convergence than the TSVD or the unfiltered approach. The filtering method additionally features a low overall additional computational complexity and appears more robust with respect to parameter selection.

In this chapter, $\|A\|_F$ will denote the Frobenius norm of A, the matrix with columns $a_1, \dots, a_m$. Vector norms without a subscript, e.g., $\|a_i\|$, will denote the vector 2-norm.

5.0.1 ▪ Background

The filtering method discussed here can be thought of as combining two approaches. The first is the angle-filtering approach of Chapter 4, which controls the angle between each column of F_k and the subspace spanned by the columns to its left (which contain more recent information). Specifically, define $\mathcal{F}_j$ as the subspace spanned by the first j columns of a given matrix F_k and define the direction sine σ_i by

$$\sigma_i = \sin(\mathcal{F}_{i-1}, f_i),\ i = 2, \dots, m_k, \text{ and } \sigma_{k,min} = \min_{i=2,\dots,m_k} \sigma_i, \tag{5.0.1}$$

where f_i is column i of matrix F_k. Given a QR factorization $F_k = QR$, we have $\sigma_i = |r_{ii}|/\|f_i\|$ (see Chapter 4). This quantity can be monitored and used to filter out columns of F_k at each iteration k.

The second approach is that of [5, 123], which sequentially drops the oldest columns of F_k, updating the QR decomposition and checking the condition number until the condition is within a given bound. This method does control the condition number, and it is both effective and efficient if the poor condition is due to disparity of l^2 lengths between the columns of F_k, as occurs in the asymptotic regime. However, if the poor condition is due to a small difference in angle between subspaces spanned by the columns of F_k, it can be inefficient and harm convergence. In particular, if the first two columns are nearly aligned, as can often happen, this method necessarily reduces F_k to a single column.

The filtering strategy discussed in this chapter combines elements from both strategies to give a guaranteed bound on the condition number after at most a single update of the QR decomposition, which has negligible cost compared with a single linear solve. Meanwhile, it maintains the angle condition which is necessary for the convergence analysis of [88] (as presented in Chapter 4) for smooth problems. As we show numerically in Section 5.3, filtering is beneficial to convergence in both the preasymptotic and the asymptotic regimes.

As in [5, 123], part of this filtering method drops older information from the system, which reduces the build-up of higher-order (nonlinear) residual terms that can be seen in the analysis of [88]. Assuming the fixed-point operator is sufficiently smooth, with Lipschitz constant κ_g and Lipschitz constant of its derivative $\hat{\kappa}_g$, Theorem 4.5 proves that the step k residual w_{k+1} satisfies

$$\|w_{k+1}\| \le \|w_k\| \left\{ \theta_k((1-\beta_k) + \kappa_g\beta_k) + C_k\hat{\kappa}_g\sqrt{1-\theta_k^2} \sum_{n=k-m_{k-1}}^{k} \|w_n\| \right\}, \tag{5.0.2}$$

where at each iteration k, the quantity C_k depends on the previous m_k values of θ_j and β_j, and increases as each of the previous m_k values of $\sigma_{j,min}$ from (5.0.1) decreases. From this estimate, we observe that in a converging iteration the largest of the higher-order terms will be from older information, i.e., $\|w_k\|\|w_{k-m_{k-1}}\|$, and can be (possibly significantly) larger than $\|w_k\|^2$.

The length filtering algorithm introduced in Subsection 5.1.2 is based on columnwise bounds for the condition number of a matrix F, assuming a lower bound on its minimal direction sine between columns as defined by (5.0.1). Computing the Frobenius norm of F columnwise is both straightforward and sharp. Our bounds for $\|F^{-1}\|_F = \|R^{-1}\|_F$ for $F = QR$ are based on the two following results for general rectangular matrices.

Summing by column over the square of the bounds from Lemma 4.13 gives the following bounds on the squared l^2 norm of each column of R^{-1}.

Proposition 5.1. *Suppose the hypotheses of Lemma 4.13 after setting $A = F$, $\hat{Q} = Q$, $\hat{R} = R$. Let s_i denote column i of R^{-1}. The following bounds hold:*

$$\|s_1\|^2 = \frac{1}{\|f_1\|^2} =: b_1, \tag{5.0.3}$$

$$\|s_2\|^2 \le \frac{1}{c_s^2}\left\{ \frac{c_t^2}{\|f_1\|^2} + \frac{1}{\|f_2\|^2} \right\} =: b_2, \tag{5.0.4}$$

$$\|s_j\|^2 \le \frac{1}{c_s^2}\left(\frac{c_t^2(c_t+c_s)^{2(j-2)}}{\|f_1\|^2 c_s^{2(j-2)}} + \sum_{i=2}^{j-1}\frac{c_t^2(c_t+c_s)^{2(j-i-1)}}{\|f_i\|^2 c_s^{2(j-i)}} + \frac{1}{\|f_j\|^2}\right\} =: b_j, \;\; 3 \le j \le m. \tag{5.0.5}$$

In the length filtering algorithm, the sums of the squared norms of the first l columns of f, multiplied by the sum over b_j, $j = 1, \ldots, l$, as defined in (5.0.3)-(5.0.5), are used to bound the square of the Frobenius condition number of F, truncated to its first l columns.

5.1 ▪ Filtered AA

The filtered Anderson acceleration (FAA) algorithm presented here consists of two filtering steps, the first filtering for disparity in column lengths and the second filtering for small angles between each column and the subspace spanned by the columns to its left. The filtered acceleration algorithm is presented in Algorithm 5.1. The individual filtering routines are described in Algorithms 5.2 and 5.3. The algorithm removes columns from F_k at each iteration $k > 1$ to ensure $\mathrm{cond}_F(F_k) \le \bar{\kappa}$, where $\bar{\kappa}$ is a user-chosen upper bound on the allowable condition number.

The FAA Algorithm 5.1 requires two additional parameters in comparison to the standard AA Algorithm 1.1. The first is $\bar{\kappa}$, a user-determined upper bound on the condition of each matrix F_k used in the least-squares problem. The second is c_s, a lower bound on the allowable direction sines between each column of each matrix F_k and the subspace spanned by columns with more recent information, as defined in (5.0.1). A discussion on parameter selection is included in Subsection 5.1.1, following the angle-filtering Algorithm 5.2.

ALGORITHM 5.1. Filtered Anderson acceleration.
The algorithm starts with a fixed point update step ($k = 0$) and an AA(1) step ($k = 1$); then the filtering starts at $k = 2$.

Choose initial iterate x_0 and parameters c_s, $\bar{\kappa}$, and m
Compute w_1, set β_0, and update $x_1 = x_0 + \beta_0 w_1$ ▷ $k = 0$
Compute w_2 ▷ $k = 1$
Set $F_1 = (w_2 - w_1)$ and $E_1 = (x_1 - x_0)$
Compute γ^2 by $F_1^T F_1 \gamma^2 = F_1^T w_2$
Set $m_1 = 1$, set β_1, and update $x_2 = x_1 + \beta_1 w_2 - (E_1 + \beta_1 F_1)\gamma^2$
for $k = 2, \ldots$ **do** ▷ $k > 1$
  Compute w_{k+1}
  Update $m_k = \min\{m_{k-1} + 1, m\}$, and drop the last columns of E_{k-1},F_{k-1} if $m_{k-1} = m$
  Set $F_k = ((w_{k+1} - w_k) \;\; (F_{k-1}))$ and $E_k = ((x_k - x_{k-1}) \;\; (E_{k-1}))$
  $[E_k, F_k, m_k] =$ Length Filter $(E_k, F_k, m_k, c_s, \bar{\kappa})$
  $[E_k, F_k, Q_k, R_k, m_k] =$ Angle Filter (E_k, F_k, m_k, c_s)
  Compute γ^{k+1} by $R_k \gamma^{k+1} = Q_k^* w_{k+1}$
  Set β_k
  Update $x_{k+1} = x_k + \beta_k w_{k+1} - (E_k + \beta_k F_k)\gamma^{k+1}$
end for

Remark 5.2 (TSVD). *We compare numerical results using Algorithm 5.1 against the standard method of controlling the condition number by TSVD, as suggested in [53]. Following [31] we implement the TSVD by computing the economy QR factorization $F_k = QR$, followed by the SVD of $R = U_R \Sigma_R V_R^*$, determine the largest s such that the ratio of singular values $\sigma_1/\sigma_s < \bar{\kappa}$,* *and then restrict U_R and V_R to their first s columns and σ_R to its first s rows and columns, referred to, respectively, as U_s, V_s, and Σ_s. The solution of the least-squares problem $\gamma^{k+1} = \operatorname{argmin} \|F_k\gamma - w_{k+1}\|$ is then given by $\gamma^{k+1} = V_s\Sigma_s^{-1}U_s^* Q^* w_{k+1}$. This procedure replaces lines 5-7 of Algorithm 5.1.*

The relative advantages (and disadvantages) to the given order of the length filtering and angle-filtering steps of Algorithm 5.1 are given in Remark 5.4. This remark also presents some minor modifications to the length filtering step that should be made if the reverse order is chosen.

5.1.1 ▪ Angle filtering

The angle-filtering algorithm was proposed by the authors in [88] (see also Chapter 4, where it is presented as Algorithm 4.1) in order to establish a one-step residual bound for smooth problems, free from an assumed l_∞ bound on the optimization coefficients. Here we reformulate Algorithm 4.1 as a module of the complete filtering strategy of Algorithm 5.1.

ALGORITHM 5.2. E, F, Q, R, m = **Angle filtering(**E, F, m, c_s**).**
Filter out columns of F by direction sine.

1: Compute the economy QR decomposition $F = QR$
2: Compute $\sigma_i = |r_{ii}|/\|f_i\|$, $i = 2, \ldots, m$, where f_i is column i of F, and r_{ii} is the corresponding diagonal entry of R
3: Remove any columns i of E and F for which $\sigma_i < c_s$
4: Update m with the new number of columns of F
5: **if** any columns were removed **then**
6: Recompute $F = QR$
7: **end if**

As mentioned above, Algorithm 5.1 has two additional parameters to choose in comparison to Algorithm 1.1. It is useful to state some heuristic advice on choosing those parameters, starting with c_s, the minimum allowable direction sine between columns of F_k. The condition estimate in the length-filtering algorithm is less sharp for smaller values of c_s. However, if c_s is chosen too large, the angle-filtering Algorithm 5.2 can reduce the algorithm to standard AA with $m = 1$ (see Subsection 5.3.4 below). We generally found a range for c_s between 0.1 and $1/\sqrt{2}$ to work well. In problems where the solution progresses through distinct preasymptotic and asymptotic regimes, as shown below in Subsection 5.3.5, it can also be advantageous to start with a higher value of c_s and dynamically decrease it to a lower value. This makes sense because a more stringent angle-filtering condition can be advantageous in the preasymptotic regime, as it prevents the build-up of higher-order residual terms. However, in the asymptotic regime, a smaller value of c_s may be preferred because the more recent information is kept, and older columns in F_k with orders of magnitude difference in length from the first are discarded. Our numerical examples show that the filtering Algorithm 5.1 also decreases the sensitivity of AA to the choice of maximum algorithmic depth m, as m can be chosen quite large, and extra columns will simply be (length) filtered out.

5.1.2 ▪ Length-filtering algorithm

The length-filtering strategy of Algorithm 5.1, as first presented in [89] is motivated by the bounds on the columns of R^{-1} from Proposition 5.1. In this algorithm, the Frobenius norm of the first k columns of F multiplied by the sum over the bounds of the squared norms of each column of R^{-1}, denoted by b_j, $j = 1, \ldots, k$, as defined in (5.0.3)-(5.0.5), are used to bound the square of the Frobenius condition number of F, denoted in the algorithm as C_F. From m_k down to 1, the constant C_F determined by the first k columns of F is updated, until it is less than a desired (user-chosen) number. On terminating this loop with some number of $\bar{k} \geq 1$ columns remaining, the columns $j > \bar{k}$ are removed from the matrices E and F.

The combination of length and angle filtering is further justified in the next section by Theorem 5.3, which showed that Algorithm 5.1 is guaranteed to control the condition of the least-squares matrix F_k at each iteration k of FAA Algorithm 5.1.

ALGORITHM 5.3. E, F, m = **Length filtering(**E, F, m, c_s**).**
Filter out columns of F by length.

1: Define $c_t = \sqrt{1 - c_s^2}$
2: Compute the vector norms $\|f_j\|^2$ and the bounds b_j for $\|s_j\|^2$, given by (5.0.3)-(5.0.5), for $j = 1, \ldots, m$
3: **for** $k = m$ down to 1 **do**
4: Compute $C_F = \left(\sum_{j=1}^k \|f_j\|^2\right)\left(\sum_{j=1}^k b_j\right)$
5: Exit loop if $C_F \leq \bar{\kappa}^2$
6: **end for**
7: $m \leftarrow k$, $E \leftarrow E(:, 1:m)$, $F \leftarrow F(:, 1:m)$

5.2 ▪ Theory

Together, as shown below in Theorem 5.3, the two low-cost filtering strategies, Algorithms 5.2 and 5.3, bound the Frobenius condition number of each least-squares matrix that arises in AA. Let $F = QR$ be an economy QR decomposition of $n \times m$ matrix F. Then $\|R\|_F = \|F\|_F$, and $\|R\|_2 = \|F\|_2$. By the equivalence $\|F\|_2 \leq \|F\|_F \leq \sqrt{m}\,\|F\|_2$ [65, Chapter 2], controlling the Frobenius condition number of F controls the 2-norm condition number of F.

We are concerned with the case $n \gg m$, meaning the number of dof is some orders of magnitude larger than the algorithmic depth $m = m_k$. In this setting, an additional computational cost of Algorithm 5.3 is negligible, and the savings of a single iteration more than make up for the additional computation.

Theorem 5.3. *Suppose the FAA Algorithm 5.1 is run with a given* $0 < c_s < 1$ *and* $\bar{\kappa} > 1$. *Then, at each iteration k, the matrix* F_k *of the least-squares problem solved in step 7 of the FAA Algorithm 5.1 by the QR decomposition has Frobenius and hence* 2*-norm condition number less than* $\bar{\kappa}$. *Moreover, both filtering steps are guaranteed to keep the first column of* F_k, *preserving the use of the most recent information.*

Proof. By the standard inequality $\|F\|_2 \leq \|F\|_F$ [65, Chapter 2], it is sufficient to establish the result for the Frobenius condition number. First we have

$$\mathrm{cond}(F)_F^2 = \|F\|_F^2 \left\|F^{-1}\right\|_F^2 = \|F\|_F^2 \left\|R^{-1}\right\|_F^2 = \left(\sum_{i=1}^m \|f_i\|^2\right) \left\|R^{-1}\right\|_F^2, \tag{5.2.1}$$

where m is the number of columns of F.

In Algorithm 5.1, the length filtering Algorithm 5.3 is done first. Since the bounds on column j of R^{-1} in Proposition 5.1 depend only on information with indices $p \leq j$, removing columns p of F for which $p > j$ does not change the column sum of index j. Hence the matrix $\hat{F}$ with $\hat{m}$ columns returned from length-filtering Algorithm 5.3 and sent to angle-filtering Algorithm 5.2 would have condition number less than $\bar{\kappa}$ if it satisfied the angle conditions $\sin(f_i, \mathcal{F}_{i-1})$,

$i = 2, \ldots, \hat{m}$. The first index $i = 1$ is not included in the previous statement, as the angle condition is satisfied on the first column by default. Moreover, Algorithm 5.3 will never remove the first column of F so long as $\bar{\kappa} \geq 1$ as $k = 1$ yields $C_F = 1$. The angle-filtering Algorithm 5.2 may remove any of columns $2, \ldots, \hat{m}$ of $\hat{F}$, which would only decrease the bounds in Proposition 5.1 by removing terms. Hence the condition estimate holds. □

Remark 5.4. *If the angle-filtering algorithm is done first, then the results of Theorem 5.3 still hold. In this case, the hypotheses of Lemma 4.13 are satisfied, and by* (5.2.1) *and Proposition 5.1, the length-filtering Algorithm 5.3 produces an output matrix with* $\mathrm{cond}(F)_F < \bar{\kappa}$. *To update the* QR *after length filtering, the QR factors can simply be truncated to agree with the updated* F. *In particular*

$$R = \begin{pmatrix} R_{11} & R_{12} \\ & R_{22} \end{pmatrix} \quad \text{and} \quad Q = \begin{pmatrix} Q_1 & Q_2 \end{pmatrix},$$

where R_{11} *and* Q_1 *correspond to the first* k *columns of* F *which are kept by length-filtering Algorithm 5.3. Then it suffices to replace* R *by* R_{11} *and* Q *by* Q_1 *(or* $Q^* w_{k+1}$ *by the corresponding entries of* $Q_1^* w_{k+1}$ *if* Q *itself is not stored).*

Although the two filtering steps could be run in either order, we run the length filtering first because it has lower computational complexity and potentially reduces the number of columns used in the QR factorization(s) of the angle-filtering Algorithm 5.2. If a sharper condition estimate is preferred, the angle filtering could be run first, after which parameter c_s could be replaced by c_s', the smallest realized value of $\sin(f_i, \mathcal{F}_{i-1}) = |r_{ii}|/\|f_i\|$, where i indexes over the columns remaining after angle filtering. Since $c_s' \geq c_s$, the result is a sharper estimate of the condition from length filtering. In our tests (not shown), this reordering did not make substantial impact on the total iterations to convergence.

5.3 ▪ Numerical experiments

In the following numerical tests, the FAA Algorithm 5.1 is tested against Anderson with the TSVD applied to the least-squares problem as in Remark 5.2. The first three examples show the advantages of filtering in more complex applications. The last two examples give insight into how filtering works in the preasymptotic vs. asymptotic regimes. In particular, as seen in Subsection 5.3.4, the length filtering dominates in the asymptotic regime, where the condition of the least-squares matrix can be compromised by the disparity in the column lengths as the algorithm converges. The angle filtering, on the other hand, stabilizes the algorithm throughout the preasymptotic regime and may be an important tool in convergence from poor initial iterates.

5.3.1 ▪ Nonlinear Helmholtz equation

Our first test is for the Picard iteration for NLH, following the problem setup and discretization described in Section 3.2. We choose parameters $k_0 = 8$ and constant $\epsilon = 0.2$. Our computations use no relaxation ($\beta = 1$) and depth $m = 20$, and we compare both the condition number and the convergence of usual AA to FAA ($\bar{\kappa} = 10^8$ and varying $c_s = 0.1, 0.2$) and to AA with TSVD (varying tolerance 10^3 and 10^8). From Figure 5.1 at left we observe the convergence of each method, and it is clear that TSVD and FAA both have much better convergence than usual AA. We also observe that filtering with $c_s = 0.1$ provides the fastest convergence, and TSVD with both 10^3 and 10^8 tolerances and FAA with $c_s = 0.2$ all have similar convergence behavior. Somewhat larger and smaller c_s (0.05, 0.3, 0.4) were also tested, with results similar to that of $c_s = 0.2$. We observe from Figure 5.1 at right that while the optimization problem of usual AA

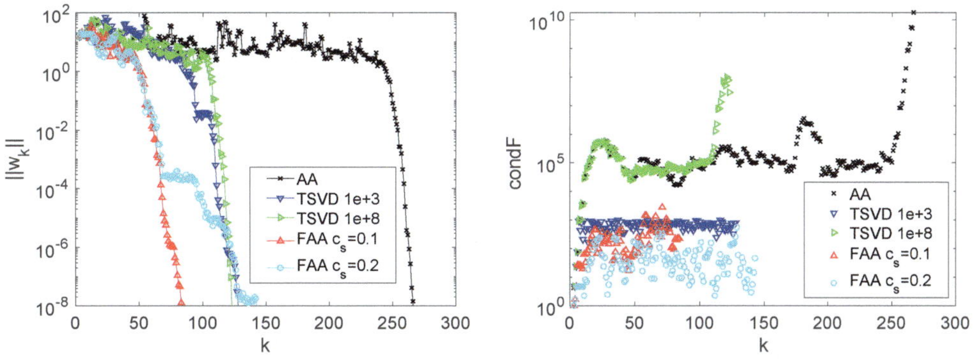

Figure 5.1. *Shown above are convergence plots (left) and condition number plots (right) for AA with* $m = 20$ *applied to the nonlinear Helmholtz problem with varying parameters for TSVD and FAA. Reprinted from [89], with permission from SIAM.*

has a high condition number near the end of the iteration, the other methods behave as designed: TSVD holds the condition number below its tolerance, and FAA holds the condition number lower than TSVD with tolerance 10^3.

5.3.2 ▪ 2D Navier-Stokes equations

We next test on the 2D driven cavity Navier-Stokes benchmark problem, using the setup, discretization, and Picard iteration described in Section 3.3. The specific spatial discretization is grad-div stabilized (with parameter 1) (P_2, P_1) Taylor-Hood elements on a triangulation created from a uniform 100×100 triangulation that is further refined once around the edges (within 0.1 from the boundary), which provides a total of 190,643 dof.

The Picard iteration is enhanced with $m = 50$ AA and no relaxation ($\beta = 1$). As found in [88, 98], large m (even $m = k - 1$) for steady NSE problems tends to work well and often better than smaller m when Re is large. We note that for this test, $m = 0$ gave no convergence, $m = 1$ gave slow convergence, and $m \geq 10$ gave convergence similar to $m = 50$ (tests omitted). A key advantage of the filtering strategy proposed herein is that one can choose large m and if m is "too large" on a particular iteration (large condition number of F), then m is effectively reduced in an optimal manner (with respect to the optimization problem).

Results for convergence and condition number vs. iteration are given in Figure 5.2 for usual AA, filtering with $\bar{\kappa} = 10^8$ and $c_s = 0.2$ and $1/\sqrt{2}$, and TSVD with tolerances 10^3 and 10^8. We observe that usual AA has a very large condition number near the end of the iteration and that both TSVD and FAA can control this. We observe good convergence with all methods except TSVD with tolerance 10^3, which displays some erratic behavior and converges somewhat slower than the other methods. Only filtering with $c_s = 1/\sqrt{2}$ beats usual AA, although only slightly.

5.3.3 ▪ Gross-Pitaevskii equations

For our next test, we consider the Gross-Pitaevskii equations (GPE). In addition to demonstrating how well FAA works on another important application problem, this test will show how filtering naturally works in conjunction with the dynamic depth techniques of [88]. The GPE and associated discrete Picard-projection (PP$_h$) scheme we will employ are described in Section 3.4, and we choose η=10000 for these tests.

PP$_h$ is run with AA $m = 1$ until the residual is below 10^{-2}, after which $m = 20$; no relaxation is used. This multilevel depth strategy of small m early and large m late in the iteration

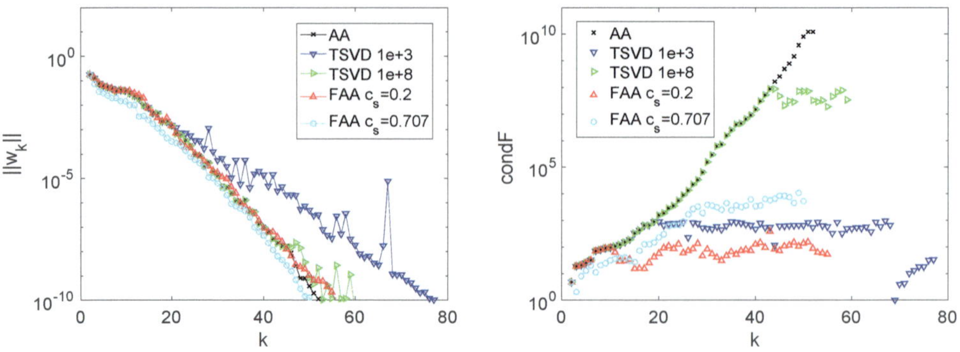

Figure 5.2. *Shown above are convergence plots (left) and condition number plots (right), for 2D Navier-Stokes driven cavity test problem with varying parameters for TSVD and filtering. Reprinted from [89], with permission from SIAM.*

proposed in [88] (and motivated by analytical convergence analysis) was found effective in [57] for this test problem, and in fact we find that if $m \geq 10$ is used for the entire iteration, then we do not get convergence (a plot of failed convergence for constant $m = 20$ AA is shown in Figure 5.3, and we note that for AA with constant $m = 20$ neither TSVD nor filtering with the same parameters as below provided for convergence).

Tests are run with multilevel $m = 1$ and then $m = 20$ AA with no filtering, FAA with $\bar{\kappa} = 10^8$ and $c_s = 0.2{:}0.1{:}0.7$, and TSVD with tolerances 10^3 and 10^8. Plots of convergence and condition numbers for these tests are shown in Figure 5.3, and we observe that usual AA, TSVD with tolerances 10^8 and 10^3, and FAA with $c_s = 0.2$ and 0.6 all had similar good convergence ($c_s = 0.3$, 0.4, and 0.5 converged similarly as well; plots are omitted). FAA with $c_s = 0.7$ converged significantly slower, and upon inspection we found that this filtering effectively reduced the AA to $m = 1$ at each step. AA with constant $m = 20$ failed to converge, with or without filtering and TSVD (at least with the same parameters as above).

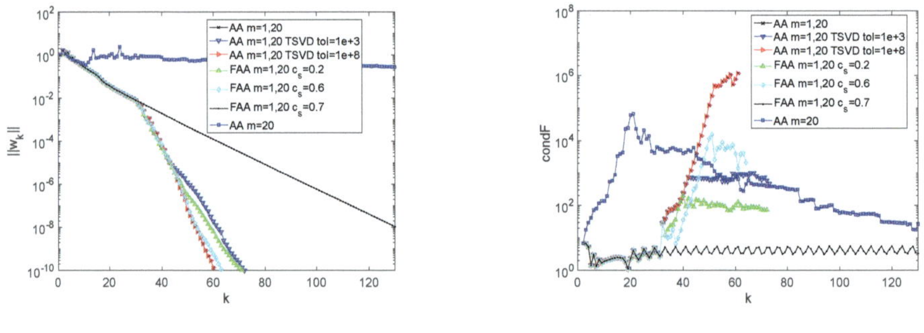

Figure 5.3. *Shown above are convergence plots (left) and condition number plots (right) for the Gross-Pitaevskii test problem. Reprinted from [89], with permission from SIAM.*

5.3.4 ▪ A monotone quasilinear equation

Next we consider a fixed-point iteration for a discretized monotone quasilinear equation, as described in Section 3.5. This relatively simple example is useful for showing the effects of varying different parameters. We use P_2 finite elements on a uniform triangularization of the unit square with right triangles and 256 subdivisions along each of the x and y axes for a total of 263,169 dof.

Table 5.1. *The number of iterations to residual convergence $\|w_k\| < 10^{-10}$ for the monotone quasilinear problem* (3.5.1) *with iteration* (3.5.5), *using FAA with $c_s = 0.1, 0.4, 2^{-1/2}$, and the TSVD with different maximum condition numbers $\bar{\kappa}$ and $m = 5, 10, 20, 40$. Without acceleration, the iteration with $\beta = \beta^*$ converges in 175 iterations, and the iteration with $\beta = 1$ does not converge. Reprinted from [89], with permission from SIAM.*

	$\beta = \beta^*$				$\beta = 1$			
FAA, $\bar{\kappa} = 10^8$	$m = 5$	$m = 10$	$m = 20$	$m = 40$	$m = 5$	$m = 10$	$m = 20$	$m = 40$
$c_s = 0.1$	32	27	27	27	21	20	20	20
$c_s = 0.4$	31	31	31	31	21	21	21	21
$c_s = 2^{-1/2}$	96	96	96	96	22	23	23	23
TSVD	$m = 5$	$m = 10$	$m = 20$	$m = 40$	$m = 5$	$m = 10$	$m = 20$	$m = 40$
$\bar{\kappa} = 10^2$	42	57	77	100	45	64	119	238
$\bar{\kappa} = 10^3$	35	45	68	96	32	38	80	164
$\bar{\kappa} = 10^4$	30	40	64	91	30	33	63	121
$\bar{\kappa} = 10^6$	33	36	58	96	30	26	45	85
$\bar{\kappa} = 10^8$	33	38	64	92	30	22	35	51

Table 5.1 shows the number of iterations to residual convergence $\|w_k\| < 10^{-10}$ from the starting guess of $u_0 = 0$. The left side of the table shows damping parameter $\beta = \beta^*$ (as given in Section 3.5) chosen to ensure the iteration is contractive. The right side shows convergence without the use of damping, for which the iteration is not contractive. With $\beta = 1$, the iteration run without AA does not converge. We compare the filtering algorithm with maximum condition number $\bar{\kappa} = 10^8$ with the TSVD algorithm with maximum condition number up to 10^8.

For the contractive iteration with $\beta = \beta^*$ we see the TSVD approach for limiting the condition number overall slows convergence as the maximum allowed condition number $\bar{\kappa}$ is decreased. The filtering approach also limits the condition number but improves convergence, so long as the minimum allowable angle c_s between columns of the least-squares matrix F_k in the angle-filtering Algorithm 5.3 is not too large. In this case, $c_s = 2^{-1/2}$ is too large and restricts the algorithmic depth after filtering to $m = 1$. Results for the FAA Algorithm 5.1 are shown here for the maximum condition number $\bar{\kappa}$ shown for the TSVD since the length-filtering Algorithm 5.3 overestimates the condition and hence as shown in the previous examples actually restricts the condition number to (sometimes significantly) less than $\bar{\kappa}$. The agreement in the iteration counts for filtering, as algorithmic depth m increasing reflects the length-filtering algorithm cutting off older columns from F_k so that increasing m does not actually change the iteration after a certain point. The TSVD continues to use all columns available, which can slow convergence.

Filtering in the contractive setting is further illustrated with Figure 5.4, which shows the columns of F_k used at each iteration k with the filtering approach with algorithmic depth $m = 10$. The left two plots illustrate the columns of F_k used for the data shown in Table 5.1 with $\bar{\kappa} = 10^8$ and $c_s = 0.1$ (leftmost) and $c_s = 0.4$ (to its right). The case with $c_s = 2^{-1/2}$ (not shown) restricts the iteration to $m = 1$ throughout the entire iteration. The right two plots in Figure 5.4 show the columns of F_k used for $c_s = 0.1$ and $c_s = 0.4$ (rightmost) with $\bar{\kappa} = 10^{16}$. For $c_s = 0.1$, the difference between the lower and higher condition bounds is that the length filtering in the second half of the iteration is suppressed for $\bar{\kappa} = 10^{16}$. This, however, does not have much effect on the total number of iterations. For $c_s = 0.4$, the angle filtering, which filters out earlier columns, dominates throughout both iterations.

Comparing the iteration counts for $c_s = 2^{-1/2}$, for $\beta = \beta^*$, and for $\beta = 1$ in Table 5.1, we see for $\beta = 1$ that the stricter requirement on the angles between columns of F_k does not make a large impact on convergence. From the rightmost plot of Figure 5.5, we see that only the second column is regularly filtered out. This illustrates that smaller damping factors can actually cause alignment of columns of F_k, which can interfere with both convergence and the condition of the least-squares problem.

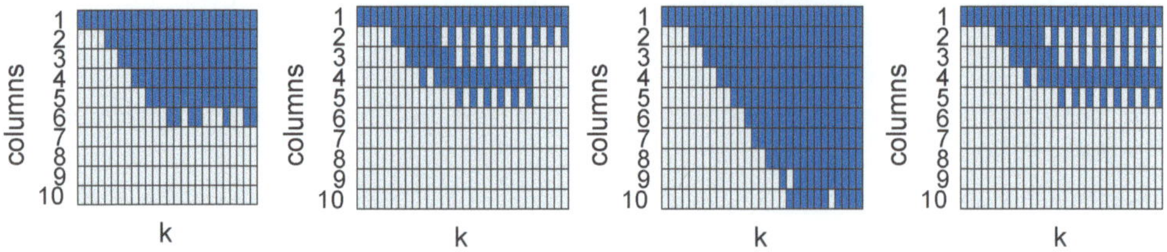

Figure 5.4. *The columns used after filtering for the monotone quasilinear problem* (3.5.1) *with iteration* (3.5.5) *with $m = 10$ and $\beta = \beta^*$. Column 1 is always used, and column 2 may be filtered out in angle filtering. From left to right: $c_s = 0.1, \bar{\kappa} = 10^8$; $c_s = 0.4, \bar{\kappa} = 10^8$ (as in Table 5.1); $c_s = 0.1, \bar{\kappa} = 10^{16}$; $c_s = 0.4, \bar{\kappa} = 10^{16}$. Reprinted from [89], with permission from SIAM.*

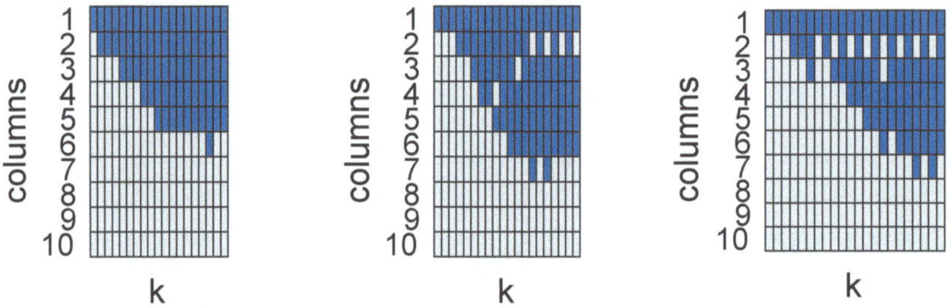

Figure 5.5. *The columns used after filtering for the monotone quasilinear problem* (3.5.1) *with iteration* (3.5.5) *with $m = 10$, $\beta = 1$, and $\bar{\kappa} = 10^8$, as in Table 5.1. Column 1 is always used, and column 2 may be filtered out in angle filtering. Left: $c_s = 0.1$, center: $c_s = 0.4$, right: $c_s = 2^{-1/2}$. Reprinted from [89], with permission from SIAM.*

Comparing with the TSVD in Table 5.1, FAA converges after fewer total iterations and is generally robust across parameter ranges tested. The TSVD shows similar performance only when the right number of columns for this problem ($m = 10$) is used and when the allowable condition number is high enough.

5.3.5 ▪ p-Laplace

In this example we consider the Picard iteration for a discretized p-Laplace equation as described in Section 3.6, using P_k Lagrange elements for $k = 1, 2, 3$, over a uniform triangularization of Ω with right triangles with 256 subdivisions in each of the x and y axes, for a total of 66,049 (P_1); 263,169 (P_2); and 591,361 (P_3) respective dof.

We compare the FAA Algorithm 5.1 with both fixed and dynamically chosen parameters to the TSVD approach in Table 5.2. The angle-filtering parameters used in this example are $c_s = 0.1$, $c_s = 2^{-1/2}$, and two dynamic choices. The first dynamic parameter definition transitions between the larger and smaller choices of c_s above based on the norm of the fixed-point update step:

$$c_s(0.1, 2^{-1/2}) = \max\left\{\min\left\{\|w_{k+1}\|^{1/2}, 2^{-1/2}\right\}, 0.1\right\}. \tag{5.3.1}$$

The second dynamic parameter definition starts with the same upper bound, $c_s \geq 2^{-1/2}$, but does not have a positive lower bound:

$$c_s(0, 2^{-1/2}) = \max\left\{\min\left\{\|w_{k+1}\|^{1/2}, 2^{-1/2}\right\}, 0\right\}. \tag{5.3.2}$$

Table 5.2. *The number of iterations to residual convergence* $\|w_k\| < 10^{-10}$ *for problem* (3.6.1) *using update* (3.6.2) *discretized with* P_k *elements with* $k = 1, 2, 3$. *Results are shown using the TSVD with different maximum condition numbers* $\bar{\kappa}$, *and FAA with* $c_s = 0.1, 2^{-1/2}$ *and a dynamically chosen* c_s *given by* (5.3.1) *and* (5.3.2). *All runs used algorithmic depth* $m = 20$.

FAA, $\bar{\kappa} = 10^8$	P_1	P_2	P_3
$c_s = 0.1$	78	98	93
$c_s = 2^{-1/2}$	443	163	262
$c_s(0.1, 2^{-1/2})$	78	84	87
$c_s(0, 2^{-1/2})$	141	81	92
TSVD	P_1	P_2	P_3
$\bar{\kappa} = 10^2$	>500	238	309
$\bar{\kappa} = 10^4$	424	169	149
$\bar{\kappa} = 10^8$	129	282	168

We will see the first definition (5.3.1) leads to lower iteration counts to convergence than (5.3.2), which demonstrates that even in the asymptotic regime, the angle condition (4.1.7) of Assumption 4.3 remains essential.

Overall, we see that the filtering, particularly with a well-chosen lower bound for c_s, allows a faster solve than the TSVD approach, while ensuring a well-conditioned least-squares problem at each step.

Chapter 6

Fitting AA theory to application problems

When I read about the evils of drinking, I gave up reading.
—Henny Youngman

The AA theory developed in Chapter 4, in particular Theorem 4.5, relies on key assumptions of the fixed-point operator. For any particular application problem and associated fixed-point operator, one can always run the AA algorithm under common heuristic guidance of low algorithmic depth, e.g., $m < 5$, and decrease the relaxation factor β until the method (hopefully) converges. However, knowing a priori what to expect is essential for robust and efficient implementations. In this chapter we show for a number of examples how to verify the hypotheses and determine the constants appearing in Theorem 4.5 for a given fixed point operator.

To discuss the assumptions, we consider seeking a fixed point of a Fréchet differentiable operator $g : X \to X$ for Hilbert space $X \subseteq \mathbb{R}^n$ equipped with inner product $(\cdot\,,\,\cdot)$ and induced norm $\|\cdot\|$. The first assumption on g is Assumption 4.1, and it assumes the following:

1. $g \in C^1(X)$ has a fixed point x^* in X.

2. There is a constant $\kappa_g > 0$ such that

$$\|g'(x)z\| \leq \kappa_g \|z\|$$

   for all $x, z \in X$.

3. There exists a constant $\hat{\kappa}_g$ such that for all $x, y, z \in X$,

$$\|g'(x)z - g'(y)z\| \leq \hat{\kappa}_g \|x - y\| \|z\| .$$

The second assumption, Assumption 4.2, states that stage-j iterates and residuals satisfy the relationship

$$\|w_{j+1} - w_j\| \geq \sigma \|e_j\| . \tag{6.0.1}$$

For many application problems, it may be possible to determine κ_g and $\hat{\kappa}_g$ and to prove that g is Lipschitz Fréchet continuously differentiable, thus satisfying Assumption 4.1. Determining σ to satisfy Assumption 4.2 can be more challenging, and details of this are discussed after Assumption 4.2 in Chapter 4. In short, in the case of contractive g (which for steady PDEs may occur when problem data is sufficiently small), i.e. when $\kappa_g < 1$, one can calculate $\sigma = 1 - \kappa_g$ [88]. If g is not contractive globally, then locally such a σ exists if the Jacobian of g does not degenerate in the vicinity of a solution. Violations of (6.0.1) can be algorithmically safeguarded against with a restarting strategy as is discussed in Chapter 4.

In this chapter, we look at solvers for three application problems: Picard iterations for the Navier-Stokes equations, Boussinesq equations, and regularized Bingham equations. For the Navier-Stokes case, we consider the mathematical analysis of this system and show that the associated fixed-point operator satisfies Assumptions 4.1 and 4.2 so that Theorem 4.5 can be invoked. Such analysis sheds light on how to best apply AA to a method and can be useful to answer questions such as whether to use large m or small m, whether the constants are mesh independent, how to choose the relaxation parameter, what the appropriate norm for the AA optimization is, and so on. For Boussinesq and regularized Bingham, we will state the results found in [92] and [90], respectively, but without proofs (which can be found in the papers) since the proofs are similar to those for Navier-Stokes but with additional details of a technical nature.

6.1 ▪ Navier-Stokes equations

We consider now properties of the Picard solver for the Navier-Stokes system, in particular of the corresponding fixed-point operator. The analysis in this section follows work from [91]. We will use the notation in this section that $b : X \times X \times X \to \mathbb{R}$ is the skew-symmetric operator defined by

$$b(v, w, z) := (v \cdot \nabla w, z) + \frac{1}{2}((\nabla \cdot v)w, z).$$

If divergence-free elements are used, then the second term drops. A key property of b is that $b(v, z, z) = 0$ for all $v, z \in X$. We will utilize the following well-known bound: there exists a constant M dependent only on the size of Ω which satisfies [80, 35]

$$b(v, w, z) \leq M\|\nabla v\|\|\nabla w\|\|\nabla z\|. \tag{6.1.1}$$

The Picard iteration for the discretized steady NSE with $f \in H^{-1}(\Omega)$ is the following: given $u_{k-1} \in V_h$, find $u_k \in V_h$ satisfying for all $v \in V_h$,

$$b(u_{k-1}, \nabla u_k, v) + \nu(\nabla u_k, \nabla v) = \langle f, v\rangle, \tag{6.1.2}$$

where we take $\nu = Re^{-1}$ to be the kinematic viscosity, $\langle\cdot,\cdot\rangle$ represents a duality pairing, and we recall from Section 3.3 that V_h is the discretely divergence-free subspace of a velocity space X_h, with X_h and the pressure space Q_h satisfying an inf-sup condition.

In order to analyze the effect of Anderson acceleration on the steady NSE Picard iteration, we next define a solution operator for the discretized Picard linearization of the NSE described in Chapter 3.

Definition 6.1. *Given $f \in H^{-1}(\Omega)$, define the Picard solution operator $G : V_h \to V_h$ as follows: Given $w \in V_h$, $G(w) \in V_h$ satisfies*

$$b(w, G(w), v) + \nu(\nabla G(w), \nabla v) = \langle f, v\rangle \;\forall v \in V_h. \tag{6.1.3}$$

The problem (6.1.3) is linear, and since $f \in H^{-1}(\Omega)$ is assumed, Lax-Milgram theory can be used to quickly show (6.1.3) is well-posed and thus that the solution operator G is well-defined. With this definition, step k of the Picard iteration for the NSE can be written as follows: set $u_k = G(u_{k-1})$.

By taking $v = G(w)$ in (6.1.3), the trilinear term vanishes, leaving

$$\nu\|\nabla G(w)\|^2 = \langle f, G(w)\rangle \leq \|f\|_{-1}\|\nabla G(w)\|,$$

and thus we have that for any $w \in V_h$,

$$\|\nabla G(w)\| \leq \nu^{-1}\|f\|_{-1}. \tag{6.1.4}$$

We now prove that G is Lipschitz continuously Fréchet differentiable and a contractive operator with contraction ratio $\mu = M\nu^{-2}\|f\|_{-1}$ when $\mu < 1$. Recall that $\mu < 1$ is a sufficient condition for uniqueness of the steady Navier-Stokes system [64].

Lemma 6.2. *The operator G is Lipschitz continuously (Fréchet) differentiable and for any $w \in V_h$ satisfies $\|\nabla G'(w)\| \leq \mu$.*

Proof. For $w,\ h \in V_h$, consider equations for $G(w)$ and $G(w+h)$ defined by (6.1.3):

$$\begin{aligned} b(w, G(w), v) + \nu(\nabla G(w), \nabla v) &= \langle f, v\rangle \ \forall v \in V_h, \\ b(w+h, G(w+h), v) + \nu(\nabla G(w+h), \nabla v) &= \langle f, v\rangle \ \forall v \in V_h. \end{aligned}$$

Subtracting yields

$$b(w+h, G(w+h) - G(w), v) + b(h, G(w), v) + \nu(\nabla(G(w+h) - G(w)), \nabla v) = 0. \tag{6.1.5}$$

Now setting $v = G(w+h) - G(w)$ vanishes the first nonlinear term, producing

$$\begin{aligned} \nu\|\nabla(G(w+h) - G(w))\|^2 &\leq |b(h, G(w), G(w+h) - G(w))| \\ &\leq M\|\nabla h\|\nabla G(w)\|\|\nabla(G(w+h) - G(w))\| \\ &\leq \nu^{-1}M\|f\|_{-1}\|\nabla h\|\|\nabla(G(w+h) - G(w))\|, \end{aligned}$$

thanks to the bounds (6.1.1) and (6.1.4). This reduces immediately to

$$\|\nabla(G(w+h) - G(w))\| \leq \mu\|\nabla h\|, \tag{6.1.6}$$

which proves G is Lipschitz continuous and contractive with contraction ratio μ.

Next we show the G is Fréchet differentiable. First define for a given $w \in V_h$ an operator $G'(w,\cdot) : V_h \to V_h$ such that for all $h \in V_h$,

$$b(h, G(w), v) + b(w, G'(w; h), v) + \nu(\nabla G'(w; h), \nabla v) = 0 \ \ \forall v \in V_h. \tag{6.1.7}$$

Using properties for G and b established above together with Lax-Milgram theory, it is easily verified that this linear problem is well-posed and thus G' is well-defined.

Subtracting (6.1.7) from (6.1.5) provides

$$\begin{aligned} b(w, G(w+h) - G(w) - G'(w; h), v) &+ \nu(\nabla(G(w+h) - G(w) - G'(w; h)), \nabla v) \\ &= -b(h, G(w+h) - G(w), v) \\ &\leq M\|\nabla h\|\|\nabla(G(w+h) - G(w))\|\|\nabla v\| \\ &\leq \mu M\|\nabla h\|^2\|\nabla v\| \end{aligned}$$

for all $v \in V_h$ thanks to (6.1.1) for the first inequality and (6.1.6) for the second. This proves that $G'(w,\cdot)$ is the Fréchet derivative of G at w. From (6.1.6) and noting $w \in V_h$ is arbitrary establishes the result. □

6.1.1 ▪ Accelerating the Navier-Stokes Picard iteration

We have now established that the associated fixed-point operator associated with the Navier-Stokes Picard iteration satisfies Assumption 4.1. Assumption 4.2 is satisfied if the problem data satisfies $\mu < 1$. Thus we can now invoke the AA theory of Theorem 4.5 for the Navier-Stokes Picard iteration.

Theorem 6.3 (Convergence result for AAPicard for Navier-Stokes Picard iteration). *Under the assumptions that the problem data satisfies $\mu < 1$ (or if Assumption 4.2 is otherwise satisfied), the kth step residual w_k generated from applying AA to the discretized Picard iteration* (6.1.2) *satisfies*

$$\|\nabla w_{k+1}\| \leq \|\nabla w_k\| \Bigg(\xi_k(1-\beta_k+\mu\beta_k) + \frac{C\mu\sqrt{1-\xi_k^2}}{2}\|\nabla w_k\| \Bigg(h(\xi_k) + 2\sum_{j=k-m_k+1}^{k-1} (k-j)\|\nabla w_j\| h(\xi_j) + m_{k-1}\|\nabla w_{k-m_k}\| h(\xi_{k-m_k}) \Bigg)\Bigg),$$

where $\xi_k := \|\nabla w_k^\alpha\| / \|\nabla w_k\|$ is the gain of the AA optimization problem as in (1.2.7), *$h(\xi_j) \leq C\sqrt{1-\xi_j^2} + \beta_j \xi_j$, and C depends on a lower bound of the direction sine between $(w_{j+1}-w_j)$ and span$\{(w_{n+1}-w_n)\}_{n=j-m_k+1}^{j-1}$.*

Remark 6.4. *In Theorem 6.3 and the other theorems in this chapter, we change the notation for the gain of the AA optimization problem defined in* (1.2.7) *to be ξ_k instead of θ_k. We do this to avoid confusion as we use θ as the temperature variable in the Boussinesq equations in the next section.*

6.2 ▪ Boussinesq equations

We next consider now the Picard solver for the steady Boussinesq system. We will continue the use of notation from the Navier-Stokes section above and also use the notation $b^* : X \times W \times W \to \mathbb{R}$ for the skew-symmetric operator defined by

$$b^*(v,s,\chi) := (v\cdot\nabla s, \chi) + \frac{1}{2}((\nabla\cdot v)s, \chi).$$

If divergence-free elements are used, then the second term drops. Similar to b, it holds for b^* that [80, 35]

$$b^*(v,s,\chi) \leq M\|\nabla v\|\|\nabla s\|\|\nabla\chi\|. \tag{6.2.1}$$

We will employ the finite element space $W_h \subset W$ for temperature/density and use V_h as the discretely divergence-free finite element space.

The discrete stationary Boussinesq equations for $(u,\theta) \in (V_h, W_h)$ can be written in FEM form as follows: Given $f \in H^{-1}(\Omega)$, $\gamma \in H^{-1}(\Omega)$, find $(u,\theta) \in (V_h, W_h)$ satisfying

$$b(u,u,v) + \nu(\nabla u, \nabla v) = Ri((0;\theta), v) + \langle f, v\rangle, \tag{6.2.2}$$
$$b^*(u,\theta,\chi) + \kappa(\nabla\theta, \nabla\chi) = \langle\gamma,\chi\rangle \tag{6.2.3}$$

for any $(v,\chi) \in (V_h, W_h)$.

The discretized Picard iteration we consider is written as follows: Given u_0, θ_0, for $k = 1, 2, 3, \ldots$, find $(u_k, \theta_k) \in (V_h, W_h)$ satisfying for any $v \in V_h, \chi \in W_h$,

$$b(u_{k-1}, u_k, v) + \nu(\nabla u_k, \nabla v) = Ri\,((0; \theta_k), v) + \langle f, v\rangle, \tag{6.2.4}$$

$$b^*(u_{k-1}, \theta_k, \chi) + \kappa(\nabla \theta_k, \nabla \chi) = \langle \gamma, \chi\rangle. \tag{6.2.5}$$

We now consider the Picard iteration in a slightly more abstract manner, in order to study the associated fixed-point function. It reads as follows: find $(\tilde{u}, \tilde{\theta}) \in (V_h, W_h)$ satisfying

$$b(u, \tilde{u}, v) + \nu(\nabla \tilde{u}, \nabla v) = Ri\left((0; \tilde{\theta}), v\right) + \langle f, v\rangle, \tag{6.2.6}$$

$$b^*(u, \tilde{\theta}, \chi) + \kappa(\nabla \tilde{\theta}, \nabla \chi) = \langle \gamma, \chi\rangle \tag{6.2.7}$$

for any $v \in V_h, \chi \in W_h$.

We have the following well-posedness and stability result from [91].

Lemma 6.5. *For $f \in H^{-1}(\Omega)^d$ and $\gamma \in H^{-1}(\Omega)$, the system* (6.2.6)-(6.2.7) *is well-posed, and solutions satisfy the bounds*

$$\|\nabla \tilde{u}\| \le K_1 \quad \text{and} \quad \|\nabla \tilde{\theta}\| \le K_2, \tag{6.2.8}$$

where K_1, K_2 are data dependent constants.

Since the iteration is well-posed, we can define its solution operator.

Definition 6.6. *Define $G_B : (V_h, W_h) \to (V_h, W_h)$ to be the solution operator of* (6.2.6)-(6.2.7). *That is,*

$$(\tilde{u}, \tilde{\theta}) = G_B(u, \theta).$$

By Lemma 6.5, we know that G_B is well-defined. The Boussinesq fixed-point iteration (6.2.4)-(6.2.5) can now be written as

$$(u_k, \theta_k) = G_B(u_{k-1}, \theta_{k-1}).$$

Definition 6.7. *Define the norm $\|(\cdot,\cdot)\|_B : (V_h, W_h) \to \mathbb{R}$ by*

$$\|(v, w)\|_B := \sqrt{\nu\|\nabla v\|^2 + \kappa\|\nabla w\|^2}.$$

The weights used in the norm definition come from the natural norm of the Boussinesq system, and this norm will be referred to as the B-norm. This weighted norm is used since it arises naturally out of the analysis, and our analysis suggests it is the natural norm to use for the optimization step of the accelerated algorithm.

The following results are proven in [92]. The first result is that G_B is Lipschitz continuous with constant dependent on the problem data.

Lemma 6.8. *There exists a positive constant C_G such that*

$$\|G_B(u, \theta) - G_B(w, z)\|_B \le C_G\|(u, \theta) - (w, z)\|_B$$

for any $(u, \theta), (w, z) \in (V_h, W_h)$. The constant C_G is defined by

$$C_G = \nu^{-1/2}\kappa^{-1/2} M\sqrt{2K_1^2 + 3K_2^2}.$$

This shows that Assumption 4.2 is satisfied under the small data restriction that yields $C_G < 1$. In our numerical tests, which used data far larger than this restriction, the σ_k's calculated at each iteration were in general no smaller than 10^{-3}. Hence it may be possible to prove that Assumption 4.2 holds under less restrictive data restrictions.

The next step is to show that G_B is Lipschitz Fréchet differentiable. To do this, we first define a mapping G'_B and then in Lemma 6.10 confirm that it is indeed the Fréchet derivative operator of G_B.

Definition 6.9. *Given $(u, \theta) \in (V_h, W_h)$, define an operator $G'_B(u, \theta; \cdot, \cdot) : (V_h, W_h) \to (V_h, W_h)$ by*

$$G'_B(u, \theta; h, s) := \left(G_B^{1'}(u, \theta; h, s), G_B^{2'}(u, \theta; h, s) \right),$$

satisfying for all $(h, s), (v, \chi) \in (V_h, W_h)$,

$$b(h, G_B^1(u, \theta), v) + b(u, G_B^{1'}(u, \theta; h, s), v) + \nu(\nabla G_B^{1'}(u, \theta; h, s), \nabla v) \\ = Ri((0; G_B^{2'}(u, \theta; h, s)), v), \tag{6.2.9}$$

$$b^*(h, G_B^2(u, \theta), \chi) + b^*(u, G_B^{2'}(u, \theta; h, s), \chi) + \kappa(\nabla G_B^{2'}(u, \theta; h, s), \nabla \chi) = 0. \tag{6.2.10}$$

Once it is established that G'_B is well-defined and is the Fréchet derivative of G_B, it follows that $G'_B(u, \theta; \cdot, \cdot)$ is the Jacobian matrix of G_B at (u, θ). From the partially decoupled system (6.2.9)-(6.2.10), it is clear that $G'_B(u, \theta; \cdot, \cdot)$ is a block upper triangular matrix. Applied to any $(h, s) \in (V_h, W_h)$, the resulting $G'_B(u, \theta; h, s)$ can be written componentwise as $\left(G_B^{1'}(u, \theta; h, s), G_B^{2'}(u, \theta; h, s) \right) \in (V_h, W_h)$.

Lemma 6.10. *The Boussinesq operator G_B is Lipschitz Fréchet differentiable: there exists a constant $\hat{C}_G$ such that for all (u, θ), (w, z), $(h, s) \in (V_h, W_h)$,*

$$\|G'_B(u, \theta; h, s)\|_B \leq C_G \|(h, s)\|_B, \tag{6.2.11}$$

and

$$\|G'_B(u + h, \theta + s; w, z) - G'_B(u, \theta; w, z)\|_B \leq \hat{C}_G \|(h, s)\|_B \|(w, z)\|_B, \tag{6.2.12}$$

where C_G is defined in Lemma 6.8 and $\hat{C}_G = 2\nu^{-1}\kappa^{-3/2} M^2 (2K_1^2 + 9K_2^2)^{1/2}$.

Proof. The proof can be found in [92]. □

6.2.1 ▪ Accelerating the decoupled Boussinesq iteration

It is now established that the associated fixed-point operator G_B satisfies Assumption 4.1 and also Assumption 4.2 for sufficiently small data, and thus we can invoke the AA theory of Theorem 4.5.

Theorem 6.11 (Convergence result for AAPicard for Boussinesq). *Under the assumptions on the data that yield $C_G < 1$ (or if Assumption 4.2 is satisfied in another way), the kth step residual (w_k, z_k) generated from applying AA to the discretized Picard iteration* (6.2.4)-(6.2.5) *satisfies*

$$\|(w_{k+1}, z_{k+1})\|_B \le \|(w_k, z_k)\|_B \Bigg(\xi_k(1 - \beta_k + C_g\beta_k) + \frac{C\hat{C}_G\sqrt{1-\xi_k^2}}{2} \Bigg(\|(w_k, z_k)\|_B h(\xi_k)$$

$$+ 2 \sum_{j=k-m_k+1}^{k-1} (k-j)\|(w_j, z_j)\|_B h(\xi_j) + m_{k-1}\|(w_{k-m_k}, z_{k-m_k})\|_B h(\xi_{k-m_k}) \Bigg) \Bigg),$$

where $\xi_k := \|(w_k^\alpha, z_k^\alpha)\|_B / \|(w_k, z_k)\|_B$ *is the gain of the AA optimization problem as in* (1.2.7), $h(\xi_j) \le C\sqrt{1-\xi_j^2} + \beta_j\xi_j$, *and* C *depends on a lower bound of the direction sine between* $(w_{j+1}, z_{j+1}) - (w_j, z_j)$ *and span*$\{(w_{n+1}, z_{n+1}) - (w_n, z_n)\}_{n=j-m_k+1}^{j-1}$.

6.3 ▪ Bingham equations

We lastly consider analysis of the AAPicard iteration for the Bercovier-Engelman regularization of the Bingham equations, which are described in Section 3.7. The FEM scheme for regularized Bingham equations is discussed in Section 3.7 and reads as follows: Given $f \in H^{-1}(\Omega)$ and $\epsilon > 0$, find $u \in V_h$ such that for all $v \in V_h$,

$$a_\epsilon(u, v) = 2\eta(Du, Dv) + \tau_s \left(\frac{Du}{|Du|_\varepsilon}, Dv \right) = \langle f, v_h \rangle. \tag{6.3.1}$$

This system is known to be well-posed [8].

The Picard iteration for (6.3.1) is given by the following: Given $u^0 \in V_h$, for $k = 1, 2, \dots,$ find $u_k \in V_h$ such that

$$2\eta(Du_k, Dv) + \tau_s \left(\frac{Du_k}{|Du_{k-1}|_\varepsilon}, Dv \right) = \langle f, v \rangle \quad \forall v \in V_h. \tag{6.3.2}$$

The convergence analysis of this iteration is given in [8]. Whether the convergence is linear (with varying linear regimes that does not approach a limit until the residual is very small) or sublinear is an open question, but from convergence plots in the literature it seems one could reasonably assume either of these might be true.

Consider properties of the solution operator of the linearized problem of the form of (6.3.2). Let $f \in H^{-1}(\Omega)$ and $u \in V_h$ be given. Consider the problem of finding $\tilde{u} \in V_h$ such that

$$2\mu(D\tilde{u}, Dv) + \tau_s \left(\frac{D\tilde{u}}{|Du|_\varepsilon}, Dv \right) = \langle f, v \rangle \quad \forall v \in V_h. \tag{6.3.3}$$

Following the same steps as in [8] for the continuous case analogue, this discrete system can also be shown to be well-posed [90].

Lemma 6.12. *For* $f \in H^{-1}(\Omega)$, $\epsilon > 0$, *and* $u \in V_h$, *(6.3.3) is well-posed and the solution satisfies the bound*

$$\|\nabla\tilde{u}\| \le \mu^{-1}\|f\|_{-1}. \tag{6.3.4}$$

Definition 6.13. *Define* $G_{RB} : V_h \to V_h$ *to be the solution operator of (6.3.3). That is,*

$$\tilde{u} = G_{RB}(u).$$

By Lemma 6.12, (6.3.3) is well-posed, and hence G_{RB} is well-defined. Thus, the iteration (6.3.2) can be written as

$$u^k = G_{RB}(u^{k-1}).$$

We now show properties of G_{RB} which are used to show convergence of the AAPicard iteration for (6.3.1) via Theorem 4.5. To prove that G_{RB} satisfies Assumption 4.1, we show that G_{RB} is Lipschitz continuous, and G'_{RB} exists and is the Fréchet derivative of G_{RB}. Next, by showing G_{RB} is Lipschitz continuously differentiable, we prove that G_{RB} satisfies the second part of Assumption 4.1. The satisfaction of both properties allows us to establish convergence of the AAPicard iteration for (6.3.2). These results are proven in [90, 8] for Lipschitz continuity and in [90] for Lipschitz continuously differentiable.

Theorem 6.14 ([90, 8]). *For any $u, w \in V_h$, it holds that*

$$\|DG_{RB}(u) - DG_{RB}(w)\| \le C_{G_{RB}} \|Dw - Du\|. \tag{6.3.5}$$

Moreover, G_{RB} is Fréchet differentiable and there exists $\hat{C}_{G_{RB}}$ satisfying for all $u, s, h \in V_h$ that

$$\|D\left(G'(u+h; s) - G'(u; s)\right)\| \le \hat{C}_{G_{RB}} \|Ds\| \|Dh\|.$$

Remark 6.15. *It remains an open problem to determine a sharp constant $C_{G_{RB}}$ that agrees with the many numerical tests that are done in [8, 90] and references therein. That is, all tests in these papers (and over a wide range of parameters) indicate that $C_{G_{RB}} \le 1$. However, the best analytical results to our knowledge are global results in [90] that scale negatively with both h and ϵ, so only for very small data and a fixed h and ϵ would the analysis imply that $C_{G_{RB}} < 1$. Similarly, in [8], $C_{G_{RB}}$ is shown to scale negatively with ϵ, and only relaxed local results can be proven to provide a contraction [8].*

Table 4.2 shows a negative effect on convergence from decreasing h, so it is also possible that the seemingly-very-pessimistic constants from [8, 90] are close to accurate in some yet-to-be-identified test cases.

6.3.1 ▪ Convergence of AAPicard for regularized Bingham equations

It is discussed above that the solution operator G of the Picard iteration (6.3.2) for regularized Bingham satisfies Assumption 4.1. To apply the one-step residual bound of [88], we also need Assumption 4.2 to be satisfied. This can come either from enforcement in the algorithm or from a small data condition (see Remark 6.15).

Theorem 6.16 (Accelerated convergence of Picard iteration for discretized regularized Bingham). *Under the assumptions on the problem data so that Assumption 4.2 is satisfied, the kth step residual w_k generated from applying AA to the discretized regularized Bingham iteration (6.3.2) satisfies*

$$\|Dw_{k+1}\| \le \|Dw_k\| \Bigg(\xi_k(1 - \beta_k + C_{G_{RB}}\beta_k) + \frac{C\hat{C}_{G_{RB}}\sqrt{1-\xi_k^2}}{2} \|Dw_k\| \Bigg(h(\xi_k) + 2 \sum_{j=k-m_k+1}^{k-1} (k-j)\|Dw_j\| h(\xi_j) + m_{k-1}\|Dw_{k-m_k}\| h(\xi_{k-m_k}) \Bigg) \Bigg),$$

where $\xi_k := \|Dw_k^{\alpha}\| / \|Dw_k\|$ is the gain of the AA optimization problem as in (1.2.7), $h(\xi_j) \le C\sqrt{1-\xi_j^2} + \beta_j \xi_j$, and C depends on a lower bound of the direction sine between $(w_{j+1} - w_j)$ and $span\{(w_{n+1} - w_n)\}_{n=j-m_k+1}^{j-1}$.

Chapter 7

Anderson-Newton

I have my standards. They're low, but I have them.
—Bette Midler

One of the most popular methods for solving nonlinear equations $f(x) = 0$ is Newton's method, $x_{k+1} = x_k - f'(x_k)^{-1}f(x_k)$, which can itself be thought of as a fixed-point iteration $x_{k+1} = g(x_k)$, with $g(x) = x - f'(x)^{-1}f(x)$. The Newton-Kantorovich theorem [83] tells us if f is smooth enough and $f'(x)$ is nonsingular at a solution x^*, then Newton's iteration converges quadratically in the vicinity of that solution. The quadratic convergence makes Newton's method very attractive. The downside of Newton's method is that outside the basin of convergence about a solution, the method can and often does exhibit dramatic failure in the form of rapid divergence. In a number of investigations including [37, 50, 92, 93, 99] we investigate the effect of applying AA to Newton iterations.

In this chapter, we consider a special case: composing AA with Newton's method applied to a singular problem, which means one for which the Jacobian $f'(x)$ is singular at a solution of $f(x) = 0$. In this circumstance, Newton iterations to find a zero of f converge linearly in the intersection of a ball and a star-shaped region about a solution [42, 43, 67, 100, 101]. In light of Chapter 4, it is natural to consider applying AA to accelerate the iteration (AANewton) .

Often PDEs depend on one or more parameters, such as the Reynolds number for NSE. A bifurcation point is a critical parameter setting which is characterized by a change in qualitative behavior of solutions. Problems become singular at bifurcations points [78, p. 8], and Newton's method in general behaves poorly both at and near bifurcation points. It is shown in [37] how Anderson acceleration can be applied to Newton iterations to improve the solve at these critical parameter points. The method developed there to ensure convergence of AANewton in the regime where the original Newton iteration converges is called *γ-safeguarded AANewton.*

It is common to numerically investigate a problem *near* rather than exactly *at* a bifurcation point, where the problem is nearly rather than exactly singular. A review of methods commonly used in this setting can be found in [38]. Sets of solutions of similar qualitative behavior are called branches; techniques for computing solution branches such as continuation [122] and deflation [22, 54] require many solves of $f(x) = 0$ as the parameter is varied, and these problems become singular or nearly singular near bifurcation points. To solve efficiently in this setting using AANewton, it is advantageous to detect whether the problem is singular and limit oneself to pure Newton iterations in the case that it is not, in order to take advantage of quadratic convergence. This is the intention behind the *adaptive γ-safeguarded* method presented in [38] and summarized below. A more complete theoretical discussion of convergence for AANewton and γ-safeguarding can be found in [37]. The main points are summarized here.

We start by defining Newton's method.

ALGORITHM 7.1. Newton's method.

Choose initial iterate x_0
for k=1,2,... **do**
    Compute w_{k+1} by $w_{k+1} = -f'(x_k)^{-1}f(x_k)$
    Update $x_{k+1} = x_k + w_{k+1}$
end for

AANewton(m), meaning AA applied to Newton's method with algorithmic depth m, is defined by computing each update $w_{k+1} = -f'(x_k)^{-1}f(x_k)$ in either Algorithm 1.1 or Algorithm 1.2. We will focus on AANewton(1), which we will refer to simply as AANewton and which is given explicitly by the following:

ALGORITHM 7.2. AANewton(1).

Choose initial iterate x_0
Set $w_1 = -f'(x_0)^{-1}f(x_0)$, and set $x_1 = x_0 + w_1$
for k=1,2,... **do**
    Compute $w_{k+1} = -f'(x_k)^{-1}f(x_k)$
    Compute $\gamma^{k+1} = (w_{k+1} - w_k)^T w_{k+1}/\|w_{k+1} - w_k\|_2^2$
    Update $x_{k+1} = x_k + w_{k+1} - \gamma^{k+1}(x_k - x_{k-1} + w_{k+1} - w_k)$
end for

7.1 ▪ One-step analysis

This section summarizes the theory developed in [37] on the effect of AA on each step in AANewton when applied to singular problems. On the one hand, the setting is quite specific to a decomposition of the error into the component in the range of $f'(x^*)$, the Jacobian of operator f at solution x^*, and its orthogonal complement, the null space of $f'(x^*)$. On the other hand, the ideas suggest some generality to how AA accelerates methods where the error can be decomposed into a slowly converging component and one that converges with higher order. Intuitively, it is not surprising that the linearly converging component in this case lies in the null space of $f'(x^*)$: if that space were trivial, then we would be back in the setting where the Jacobian is nonsingular in the vicinity of a solution, which is essentially the nondegeneracy assumption (4.1.6), under which Newton could be expected to converge quadratically.

For the theory discussed in this chapter, we take $f : \mathbb{R}^n \to \mathbb{R}^n$ to be a C^3 function, $f(x^*) = 0$, $N = \text{null}\,(f'(x^*))$, $R = \text{range}\,(f'(x^*))$, $\dim N = 1$, and P_N and P_R to be the orthogonal projections onto N and R, respectively. The assumption that $R \perp N$ is common in the singular Newton literature [42, 66], and no generality is lost in finite dimensions. Newton's method is essentially invariant under nonsingular affine transformations of the domain and nonsingular linear transformations of the range [45]. Thus, to determine the convergence behavior of Newton's method applied to a general C^3 function $F : \mathbb{R}^n \to \mathbb{R}^n$, it suffices to study that of $f(x) = U^T F(x) V$, where U and V come from the singular value decomposition $F'(x^*) = U\Sigma V^T$. Since $f'(x^*) = U^T F'(x^*) V = \Sigma$, it follows that $\text{null}\,(f'(x^*)) \perp \text{range}\,(f'(x^*))$. We take the *singular set* S to be the set of all $x \in \mathbb{R}^n$ such that $\det\big(f'(x)\big) = 0$.

For the remainder of this chapter, let $\|\cdot\|$ denote the Euclidean 2-norm, and let $B_\rho(x)$ denote a ball of radius ρ centered at x, $e_k = x_k - x^*$ (note that this is different than how e_k is defined elsewhere in this book), where x^* is a solution to $f(x) = 0$, and $w_{k+1} = -f'(x_k)^{-1}f(x_k)$. Let θ_{k+1} be defined as in (1.2.7), and let γ^{k+1} be computed as stated in Algorithm 7.2.

Let $\hat{D}_N(x) := P_N f''(x^*)(P_N(x-x^*), P_N)$. If $\hat{D}_N(x)$ is invertible as a map on N whenever $P_N(x - x^*) \neq 0$, one can show that there exist constants $\hat{\rho} > 0$ and $\hat{\sigma} > 0$ for which $f'(x)$ is invertible in the region

$$\hat{W} := W(\hat{\rho}, \hat{\sigma}, x^*) = \{x \in \mathbb{R}^n : \|x - x^*\| < \hat{\rho}, \ \|P_R(x - x^*)\| < \hat{\sigma}\|P_N(x - x^*)\|\} \quad (7.1.1)$$

and that the standard Newton iterates remain in this region for sufficiently small initial error $\|e_0\|$. Further, $f'(x)^{-1} = \hat{D}_N(x)^{-1} + \mathcal{O}(1)$ and $\|f'(x)^{-1}\| \leq c\|x - x^*\|^{-1}$ (see [42, 43], or [67] for details). For the one-step analysis, we will work under the following assumption.

Assumption 7.1. *The linear operator $\hat{D}(x)$ is invertible as a map on N in the ball $B_{\hat{r}}(x^*) \setminus S$, and $\|x - x^*\| < \hat{r}$ implies $f'(x)^{-1} = \hat{D}(x)^{-1} + \mathcal{O}(1)$.*

In [37] and summarized here in Section 7.2 , we show that with safeguarding and sufficiently small $\|e_0\|$, the AANewton iterates remain in $\hat{W}$ if $x_0 \in \hat{W}$.

Due to the technicality of the argument that follows, we include the steps of the argument, but not their proofs, which can be found in [37]. We start with a standard error expansion for the analysis of Newton's method for singular problems, and show how the behavior of AANewton iterates differ from the behavior of Newton iterates without acceleration, revealing the underlying mechanism of the acceleration.

7.1.1 ▪ Decomposition of the error

Theorem 7.2. *Let Assumption 7.1 hold, and let $x_k \in B_{\hat{r}}(x^*) \setminus S$. Then*

$$e_k + w_{k+1} = \frac{1}{2}P_N e_k + \frac{1}{2}\hat{D}^{-1}(x_k)f''(x_k)(e_k, P_R e_k) + \mathcal{O}(\|e_k\|^2), \quad (7.1.2)$$

$$w_{k+1} = -\frac{1}{2}P_N e_k + \left(\frac{1}{2}\hat{D}^{-1}(x_k)f''(x_k)(e_k, \cdot) - I\right)P_R e_k + \mathcal{O}(\|e_k\|^2). \quad (7.1.3)$$

Bounds similar to (7.1.2) may be found in [42, 43, 101]. Note that (7.1.3) follows from (7.1.2) by subtracting e_k to the right-hand side, writing $e_k = P_N e_k + P_R e_k$, and grouping like terms.

We will denote $T_k(\cdot) = (1/2)\hat{D}^{-1}(x_k)f''(x_k)(e_k, \cdot)$. Let γ_{k+1} be the coefficient computed in Algorithm 7.2. For $x, y \in \mathbb{R}^n$, define the function

$$L_{k+1}(x, y) := (1 - \gamma_{k+1})x + \gamma_{k+1}y. \quad (7.1.4)$$

We will call any term of the form $L_{k+1}(x, y)$ the *AANewton sum* of x and y. In light of (1.2.4), we have $L_{k+1}(x_k, x_{k-1}) = x_k^\alpha$. In Chapter 4, the optimization gain θ_{k+1} was shown to be the key quantity determining the acceleration from one step of AA in the nonsingular case. Here, in the singular case, it will be shown to determine the acceleration of the error along the null component.

With the notation described above, (7.1.2) and (7.1.3) become

$$e_k + w_{k+1} = \frac{1}{2}P_N e_k + T_k P_R e_k + \mathcal{O}(\|e_k\|^2), \quad (7.1.5)$$

$$w_{k+1} = -\frac{1}{2}P_N e_k + (T_k - I)P_R e_k + \mathcal{O}(\|e_k\|^2). \quad (7.1.6)$$

The following proposition provides the error expansions based on (7.1.5)-(7.1.6) that we will use in our analysis.

Proposition 7.3. *Let q_{k-1}^k denote a term for which $\|q_{k-1}^k\| \le c\big(|1-\gamma_{k+1}|\,\|e_k\|^2+|\gamma_{k+1}|\,\|e_{k-1}\|^2\big)$. Under the assumptions of Theorem 7.2, if x_{k+1} is the $(k+1)$st AANewton iterate, then we can expand e_{k+1} and w_{k+1}^α as*

$$e_{k+1} = \frac{1}{2}P_N e_k^\alpha + (T_k P_R e_k)^\alpha + q_{k-1}^k, \tag{7.1.7}$$

$$w_{k+1}^\alpha = -\frac{1}{2}P_N e_k^\alpha + ((T_k - I)P_R e_k)^\alpha + q_{k-1}^k. \tag{7.1.8}$$

We can obtain a simple upper bound on $P_R e_{k+1}$ by applying P_R to (7.1.7). By orthogonality, and noting the range of T_k lies in N, the only term remaining on the right-hand side is q_{k-1}^k. Thus

$$\|P_R e_{k+1}\| \le |1-\gamma_{k+1}|C_1\|e_k\|^2 + |\gamma_{k+1}|C_2\|e_{k-1}\|^2. \tag{7.1.9}$$

The constants C_1 and C_2 are independent of k and depend on f. The bound in (7.1.9) resembles the result of Lemma 1 in [93] for depth $m = 1$. There, the Jacobian is assumed to be nonsingular at the solution x^*. So like a standard Newton step, the range component of the error from a AANewton step behaves as if the Jacobian were nonsingular at the solution. Thus, when the Jacobian is singular at x^*, the source of slow convergence must come from the null space component. Conversely, if AANewton is seen to accelerate a given Newton sequence, and $f'(x^*)$ is singular, then it must accelerate the null component. Because of this, the next part of the analysis focuses on the null component error, which we develop based on the notion of *pair types*. This method introduces a new technique for the one-step analysis of AA and explains how the convergence rate changes at different steps for AANewton.

7.1.2 ▪ Pair types

We can consider the AANewton algorithm as acting on ordered pairs (x_k, x_{k-1}), where x_k and x_{k-1} are two consecutive AANewton iterates. We then analyze the output, x_{k+1}, based on where the vectors x_k and x_{k-1} lie in $\mathbb{R}^n$; e.g., x_k and x_{k-1} both lie near N, or x_k lies near N and x_{k-1} lies near R. This leads to the definition of pair types.

Definition 7.4. *Let $\{x_k\}$ be a sequence of AANewton iterates.*

1. *(x_k, x_{k-1}) is an* N-pair *if $(1/2)P_N e_i$ is the dominant term in (7.1.5) for $i = k, k-1$,*
2. *(x_k, x_{k-1}) is an* R-pair *if $T_i P_R e_i$ is the dominant term in (7.1.5) for $i = k, k-1$,*
3. *(x_k, x_{k-1}) is an* NR-pair *if $(1/2)P_N e_i$ is the dominant term for $i = k$, and $T_i P_R e_i$ is the dominant term for $i = k-1$, and*
4. *(x_k, x_{k-1}) is an* RN-pair *if $T_i P_R e_i$ is the dominant term for $i = k$, and $(1/2)P_N e_i$ is the dominant term for $i = k-1$.*

Each respective pair is a strong pair *if the corresponding AANewton sum is dominant in* (7.1.7).

Strong pairs are those for which the dominant terms in x_k and x_{k-1} remain dominant in x_{k+1}. Our analysis focuses on strong pairs, and of particular interest is how AANewton acts on strong N-pairs, since near N the standard Newton method exhibits linear convergence. Strong N-pairs are the most responsive to a successful optimization step in AANewton.

7.1.3 ▪ The null-space component and compatibility

We next derive a general relation between the null component error and the optimization gain θ_{k+1}. The strategy is to take the largest term (with respect to the norm) in the expansion of $P_N e_{k+1}$ from (7.1.7) and find conditions under which this term is bounded by $\theta_{k+1}\|w_{k+1}\|$. We will then see (in the next section) when each strong pair type satisfies these conditions. First, consider each term in expansion (7.1.7). Namely,

$$\tfrac{(1-\gamma_{k+1})P_N e_k}{2},\ \tfrac{\gamma_{k+1}P_N e_{k-1}}{2},\ (1-\gamma_{k+1})T_k P_R e_k,\ \gamma_{k+1}T_{k-1}P_R e_{k-1},\ \text{and } P_N q_{k-1}^k. \tag{7.1.10}$$

Any sum of a given strict subset of these terms could be the dominant term in (7.1.7). We'll let S_{k+1}^e denote the set of all such sums. For example, if (x_k, x_{k-1}) is a strong N-pair as defined in Definition 7.4, then $(1-\gamma_{k+1})P_N e_k + \gamma_{k+1}P_N e_{k-1}$ is the dominant term. We'll write

$$R_{k+1}^e := \{\|P_N e_{k+1} - \widetilde{E}_{k+1}\|/\|\widetilde{E}_{k+1}\| : \widetilde{E}_{k+1} \in S_{k+1}^e\}, \tag{7.1.11}$$

and $E_{k+1} := \operatorname{argmin} R_{k+1}^e$. We then define

$$r_{k+1}^e := \min R_{k+1}^e = \|P_N e_{k+1} - E_{k+1}\|/\|E_{k+1}\|. \tag{7.1.12}$$

Observe that if $q \in R_{k+1}^e$, then $1/q \in R_{k+1}^e$. This gives the following lemma.

Lemma 7.5. *The minimum ratio r_{k+1}^e satisfies $r_{k+1}^e \le 1$ for all $k \ge 0$.*

Analogous notation will be used for the expansion $P_N w_{k+1}^\alpha$ in (7.1.8). Namely, S_{k+1}^w will denote the set of sums of strict subsets of the terms

$$-\tfrac{(1-\gamma_{k+1})}{2}P_N e_k,\ -\tfrac{\gamma_{k+1}P_N e_{k-1}}{2},\ (1-\gamma_{k+1})T_k P_R e_k,\ \gamma_{k+1}T_{k-1}P_R e_{k-1},\ \text{and } P_N q_{k-1}^k, \tag{7.1.13}$$

We'll write $R_{k+1}^w = \{\|P_N w_{k+1}^\alpha - \widetilde{F}_{k+1}\|/\|\widetilde{F}_{k+1}\| : \widetilde{F}_{k+1} \in S_{k+1}^w\}$, $F_{k+1} := \operatorname{argmin} R_{k+1}^w$, and $r_{k+1}^w = \min R_{k+1}^w$. Next we look at bounding the null-space component of the error in terms of the optimization gain θ_{k+1}.

Proposition 7.6. *Let the assumptions of Theorem 7.2 hold for x_k and x_{k-1}. Let q_{k-1}^k denote a term for which $\|q_{k-1}^k\| \le c(|1-\gamma_{k+1}|\,\|e_k\|^2 + |\gamma_{k+1}|\,\|e_{k-1}\|^2)$, and suppose*

$$\|E_{k+1}\| \le \frac{\|P_N w_{k+1}^\alpha\|}{1 - r_{k+1}^e}. \tag{7.1.14}$$

Then

$$\|P_N e_{k+1}\| \le \left(\frac{1 + r_{k+1}^e}{1 - r_{k+1}^e}\right)\theta_{k+1}\|w_{k+1}\|. \tag{7.1.15}$$

Proposition 7.7. *The relation* (7.1.14) *holds if $r_{k+1}^e < 1$, and there is an element $\widetilde{F}_{k+1} \in S_{k+1}^w$, such that $\|\widetilde{F}_{k+1}\| = \|E_{k+1}\|$, and $\|P_N e_{k+1} - E_{k+1}\| = \|P_N w_{k+1}^\alpha - \widetilde{F}_{k+1}\|$.*

As r_{k+1}^e approaches 1, the denominator in (7.1.15) approaches zero, resulting in a poor bound. However, the following proposition shows that $r_{k+1}^e \approx 1$ implies $P_N e_{k+1}$ was accelerated.

Proposition 7.8. *Let r^e_{k+1} be defined as in* (7.1.12)*, and let the assumptions of Theorem 7.2 hold for x_k and x_{k-1}. If $0 \le \varepsilon < 1$ and $r^e_{k+1} = 1 - \varepsilon$, then $P_N e_{k+1} = ((2-\varepsilon)/(1-\varepsilon))\mathcal{O}(\max\{\|e_k\|^2, \|e_{k-1}\|^2\})$.*

Proposition 7.6 and Proposition 7.8 together broadly say that either there is a strongly dominant term E_{k+1}, so that $r^e_{k+1} \ll 1$, or there is not a strongly dominant term and $r^e_{k+1} \approx 1$, in which case we may bound $\|P_N e_{k+1}\|$ by higher-order terms. In the former case, we have $(1 + r^e_{k+1})(1 - r^e_{k+1})^{-1} \le C$, where $C > 1$ and approaches 1 as r^e_{k+1} approaches zero. This type of bound is significant and motivates Definition 7.9 below. With γ-safeguarding, introduced in Section 7.2, we can show that if x_k and x_{k-1} are sufficiently close to N and x^*, then r^e_{k+1} is bounded away from 1 and approaches zero as x_k and x_{k-1} approach N and x^*. In other words, one can say that the constant C approaches 1 as x_k and x_{k-1} approach N and x^*.

In general, if $\|P_N e_{k+1}\| \le C\theta_{k+1}\|w_{k+1}\|$, where $C > 1$ may be taken as close to 1 as desired as r^e_{k+1} approaches zero, we say that C is a *moderate constant.* We now use this idea to define a *compatible step.*

Definition 7.9. *Let $\{x_k\}$ be a sequence of AANewton iterates. We say that x_{k+1} is* compatible *or a* compatible step *if there exists a moderate constant $C > 0$ such that $\|P_N e_{k+1}\| \le C\theta_{k+1}\|w_{k+1}\|$, in which case we will write $P_N e_{k+1} = \mathcal{O}(\theta_{k+1}\|w_{k+1}\|)$. Otherwise, (x_k, x_{k-1}) is an* incompatible pair, *and x_{k+1} is* incompatible *or an* incompatible step

Compatible here relates to the result of Proposition 7.6. If (x_k, x_{k-1}) is a pair that satisfies Proposition 7.6, then $\|P_N e_{k+1}\| = \mathcal{O}(\theta_{k+1}\|w_{k+1}\|)$. Hence a successful optimization step in AANewton implies acceleration of $P_N e_{k+1}$ and therefore of e_{k+1}. When (x_k, x_{k-1}) is *incompatible,* i.e., $\|P_N e_{k+1}\| \ne \mathcal{O}(\theta_{k+1}\|w_{k+1}\|)$, acceleration may still occur. However, this is not guaranteed, and a successful optimization step does not imply acceleration of $P_N e_{k+1}$.

So far, it has been shown that the expansion of $P_N e_{k+1}$ in (7.1.7) either has a dominant term, and this term is $\mathcal{O}(\theta_{k+1}\|w_{k+1}\|)$ when (x_k, x_{k-1}) is a compatible pair, or there is no dominant term and $P_N e_{k+1} = \mathcal{O}(\max\{\|e_k\|^2, \|e_{k-1}\|^2\})$.

Next we focus on the analysis of strong pair types, using the results above. If a given pair (x_k, x_{k-1}) is not strong, and there is a dominant term in (7.1.7), then this term can be analyzed analogously to the strong terms analyzed here.

7.1.4 ▪ Compatibility conditions

Our aim here is to establish conditions for each strong pair type under which compatibility is assured, that is, conditions under which $\|P_N e_{k+1}\| = \mathcal{O}(\theta_{k+1}\|w_{k+1}\|)$. Each pair type can be compatible if certain alignment conditions are met. These vary by pair type, with strong mixed pairs having the most stringent alignment conditions. On the other hand, strong N-pairs are automatically compatible.

Lemma 7.10. *Let the assumptions in Theorem 7.2 hold for x_k and x_{k-1}, and suppose $r^e_{k+1} < 1$. If (x_k, x_{k-1}) is a strong N-pair, then (x_k, x_{k-1}) is a compatible pair.*

Remark 7.11. *Lemma 7.10 is significant because it says that if x_k and x_{k-1} are close to N, and the resulting AANewton step remains near N, the region in which standard Newton is slowest, then (x_k, x_{k-1}) is guaranteed to be compatible, and therefore $P_N e_{k+1}$ is controlled by $\theta_{k+1}\|w_{k+1}\|$. So, for θ_{k+1} sufficiently small, strong N-pairs are accelerated, and as x_k and x_{k-1} tend to N and the root x^*, $\|P_N e_{k+1}\|/\|P_N e_k\| \to \theta_{k+1}$. Should θ_{k+1} remain bounded* *away from zero, then to first order, AANewton will perform no worse and in general will perform better than standard Newton near N since $\theta_{k+1} \leq 1$. In Section 7.2 we introduce γ-safeguarding, which will ensure the iterates remain near N if x_0 is chosen near N, thereby guaranteeing compatibility.*

Lemma 7.12 gives conditions in which a strong R-pair is compatible and a worst-case bound for the incompatible case.

Lemma 7.12. *Let the assumptions in Theorem 7.2 hold for x_k and x_{k-1}, let (x_k, x_{k-1}) be a strong R-pair, and suppose $r^e_{k+1} < 1$. Then (x_k, x_{k-1}) is compatible if $(P_N e^\alpha_k)^T q^k_{k-1} \geq 0$. Otherwise, if (x_k, x_{k-1}) is incompatible, then*

$$\|P_N e_{k+1}\| \leq (1 + r^e_{k+1})\Big(|1 - \gamma_{k+1}|\, C_1\, q^{k-1}_{k-2} + |\gamma_{k+1}|\, C_2\, q^{k-2}_{k-3} \Big), \tag{7.1.16}$$

where q^i_{i-1} is a term such that $\|q^i_{i-1}\| \leq |1 - \gamma_{i+1}| C_3 \|e_i\|^2 + C_4 \mathcal{O}(\|e_{i-1}\|^2)$ for $i \in \{k-2, k-1\}$, where C_3 and C_4 are constants determined by f.

Now we come to strong mixed pairs, the case with the most stringent alignment conditions. As in Proposition 7.6, $E_{k+1} = \operatorname{argmin} R^e_{k+1}$, where R^e_{k+1} is defined in (7.1.11). We will also use the *AANewton sum* notation $L_k(x, y)$ be defined as in (7.1.4).

Lemma 7.13. *Let the assumptions of Theorem 7.2 hold for x_k and x_{k-1}, and suppose $r^e_{k+1} < 1$. Suppose (x_k, x_{k-1}) is a strong mixed pair. There are two cases.*

1. *(Strong NR-pair) If $E_{k+1} = L_{k+1}(P_N e_k, T_{k-1} P_R e_{k-1})$, then (x_k, x_{k-1}) is compatible if*

   (a) $((1 - \gamma_{k+1}) P_N e_k)^T (\gamma_{k+1} T_{k-1} P_R e_{k-1}) \leq 0$, *and*

   (b) $(\gamma_{k+1} P_N e_{k-1})^T ((1 - \gamma_{k+1}) T_k P_R e_k + q^k_{k-1}) \geq 0$.

   When (x_k, x_{k-1}) is not compatible,

   $$\|P_N e_{k+1}\| < C(1 + r^e_{k+1}) \max\{|1 - \gamma_{k+1}|\, \|P_N e_k\|, |\gamma_{k+1}|\, \|T_{k-1} P_R e_{k-1}\|\}. \tag{7.1.17}$$

2. *(Strong RN-pair) If $E_{k+1} = L_{k+1}(T_k P_R e_k, P_N e_{k-1})$, then (x_k, x_{k-1}) is compatible if*

   (a) $((1 - \gamma_{k+1}) T_k P_R e_k)^T (\gamma_{k+1} P_N e_{k-1}) \leq 0$, *and*

   (b) $((1 - \gamma_{k+1} P_N e_k)^T (\gamma_{k+1} T_{k-1} P_R e_{k-1} + q^k_{k-1}) \geq 0$.

   When (x_k, x_{k-1}) is not compatible,

   $$\|P_N e_{k+1}\| \leq C(1 + r^e_{k+1}) \max\{|1 - \gamma_{k+1}|\, \|T_k P_R e_k\|, (|\gamma_{k+1}|/2)\, \|P_N e_{k-1}\|\}. \tag{7.1.18}$$

In both cases, C denotes a constant determined by f.

The results of Lemmas 7.10, 7.12, and 7.13 are summarized in the following theorem.

Theorem 7.14. *Let the assumptions of Theorem 7.2 hold for x_k and x_{k-1}. Suppose (x_k, x_{k-1}) is a strong pair. If (x_k, x_{k-1}) is a strong N-pair, then (x_k, x_{k-1}) is compatible. Otherwise, there are alignment conditions under which (x_k, x_{k-1}) is compatible. If (x_k, x_{k-1}) is an incompatible strong R-pair, then $P_N e_{k+1}$ is bounded by higher-order terms. If (x_k, x_{k-1}) is an incompatible strong mixed pair, then $P_N e_{k+1}$ is no worse than $\mathcal{O}\big(\max\{|1 - \gamma_{k+1}|\, \|e_k\|, |\gamma_{k+1}|\, \|e_{k-1}\|\}\big)$.*

7.1.5 ▪ Incompatible pairs

Now we look at the incompatible case. We will use the quantity $\sigma_{k+1} := \|P_R e_{k+1}\| / \|P_N e_{k+1}\|$. One can think of σ_{k+1} as measuring the angle between e_{k+1} and the null space N, but it can also be interpreted as measuring the acceleration of $\|P_N e_{k+1}\|$ relative to the standard Newton algorithm. In a standard Newton step, $P_N e_{k+1} = \mathcal{O}(\|e_k\|)$, and $P_R e_{k+1} = \mathcal{O}(\|e_k\|^2)$. Hence $\sigma_{k+1} = \mathcal{O}(\|e_k\|)$. So we can think of any AANewton step as "Newton-like" when $\sigma_{k+1} = \mathcal{O}(\max\{|1-\gamma_{k+1}|\,\|e_k\|, |\gamma_{k+1}|\,\|e_{k-1}\|\})$ and as accelerated if $\sigma_{k+1} = \mathcal{O}(\max\{|1-\gamma_{k+1}|\,\|e_k\|^\ell, |\gamma_{k+1}|\,\|e_{k-1}\|^\ell\})$ with $0 \le \ell < 1$. From this perspective, the worst case for a AANewton step is when $P_N e_{k+1}$ is incompatible and bounded below by first order in the sense that $\|E_{k+1}\| > C\max\{|1-\gamma_{k+1}|\,\|e_k\|, |\gamma_{k+1}|\,\|e_{k-1}\|\}$.

Lemma 7.15. *Let the assumptions of Theorem 7.2 hold for x_k and x_{k-1}. As in Proposition 7.6, let $E_{k+1} = \operatorname{argmin} R^e_{k+1}$. If*

$$\|E_{k+1}\| > C\max\{|1-\gamma_{k+1}|\,\|e_k\|, |\gamma_{k+1}|\,\|e_{k-1}\|\}, \tag{7.1.19}$$

where constant $C > 0$ depends on f, and $1 - r^e_{k+1} \neq 0$, then $\sigma_{k+1} = \mathcal{O}(\max\{\|e_k\|, \|e_{k-1}\|\})$.

Remark 7.16. *Under the same assumptions, we can make the slightly stronger statement: Suppose $\|E_{k+1}\| > C\max\{|1-\gamma_{k+1}|\,\|e_k\|^\ell, |\gamma_{k+1}|\,\|e_{k-1}\|^\ell\}$ with $0 < \ell < 2$. Then $\sigma_{k+1} \le (C/(1-r^e_{k+1}))\max\{\|e_k\|^{2-\ell}, \|e_{k-1}\|^{2-\ell}\}$.*

By definition, $\|P_R e_{k+1}\| = \sigma_{k+1}\|P_N e_{k+1}\|$. So if $\sigma_{k+1} = \mathcal{O}(\max\{\|e_k\|, \|e_{k-1}\|\})$, then

$$\|P_R e_{k+1}\| \le C\max\{\|e_k\|, \|e_{k-1}\|\}\|P_N e_{k+1}\|. \tag{7.1.20}$$

This says that if x_{k+1} is a Newton-like step and $\max\{\|e_k\|, \|e_{k-1}\|\} \ll 1$, then $\|P_R e_{k+1}\| \ll \|P_N e_{k+1}\|$. Therefore x_{k+1} is close to N. That is, Newton-like steps send iterates towards N, the region in which compatibility is guaranteed by Lemma 7.10. Moreover, the smaller $\max\{\|e_k\|, \|e_{k-1}\|\}$ is, the closer Newton-like steps are to N. The next theorem shows that after at most a small number of Newton-like steps, the iterates will be sufficiently clustered around N to yield a compatible pair. This explains the inconsistent rates of acceleration we see in AANewton: slowly converging steps necessarily imply an upcoming accelerated step. While we see similar effects in AA under the conditions of Chapter 4, the phenomenon has yet to be clearly explained in that scenario.

Theorem 7.17. *Let Assumption 7.1 hold, and let $x_i \in B_{\hat{r}}(x^*)\backslash S$ for $i = k, k-1, k-2$, and $k-3$. Let $E_k = \operatorname{argmin} R^e_k$ and $E_{k-1} = \operatorname{argmin} R^e_{k-1}$. Suppose x_j and x_{j-1} are incompatible with $\|E_j\| > C\max\{|1-\gamma_j|\,\|e_{j-1}\|, |\gamma_j|\,\|e_{j-2}\|\}$, and r^e_j is bounded away from one for $j = k, k-1$. Then $P_N e_{k+1}$ is either $\mathcal{O}(\theta_{k+1}\|w_{k+1}\|)$, i.e., (x_k, x_{k-1}) is compatible or consists only of higher order terms.*

Proof. Apply the triangle inequality to (7.1.7) to get

$$\|P_N e_{k+1}\| \le \frac{1}{2}\|P_N e^\alpha_k\| + |1-\gamma_{k+1}|\,C\,\|P_R e_k\| + |\gamma_{k+1}|\,C\,\|P_R e_{k-1}\| + q^k_{k-1}. \tag{7.1.21}$$

By Lemma 7.15 and (7.1.20), $\|P_R e_i\| \le C\|e_i\|\max\{\|e_{i-1}\|, \|e_{i-2}\|\}$ for $i = k, k-1$. Grouping the higher-order terms into a single term denoted by Q_{k+1},[4] (7.1.21) becomes

$$\|P_N e_{k+1}\| \le \frac{1}{2}\|P_N e_k^\alpha\| + Q_{k+1}. \tag{7.1.22}$$

There are now two cases. If $(1/2)\|P_N e_k^\alpha\| > \|Q_{k+1}\|$, then

$$\begin{aligned} \|P_N e_{k+1}\| &\le \left((1 + \frac{\|Q_{k+1}\|}{(1/2)\|P_N e_k^\alpha\|}\right)(1/2)\|P_N e_k^\alpha\| \\ &\le \left(\frac{1 + \frac{\|Q_{k+1}\|}{(1/2)\|P_N e_k^\alpha\|}}{1 - \frac{\|Q_{k+1}\|}{(1/2)\|P_N e_k^\alpha\|}}\right)\theta_{k+1}\|w_{k+1}\|. \end{aligned} \tag{7.1.23}$$

The last inequality follows from applying the reverse triangle inequality to the expansion of $P_N w_{k+1}^\alpha$ from (7.1.8) as in the proof of Proposition 7.7 (see [37, Proposition 4.2]). Otherwise, if $(1/2)\|P_N e_k^\alpha\| \le \|Q_{k+1}\|$, then $\|P_N e_{k+1}\| \le 2\|Q_{k+1}\|$, so that $\|P_N e_{k+1}\|$ is bounded only by higher-order terms. □

From Lemma 7.13, incompatible pairs do not necessarily lead to improved error. However, we are assured by Theorem 7.17 that no more than two such steps may occur, at which point we can expect improved error (locally) either by compatibility, for sufficiently small θ_{k+1}, or from higher-order terms dominating.

In the next section we introduce the safeguarding strategy called γ-safeguarding, under which we can show convergence.

7.2 ▪ The γ-safeguarding algorithms

A major challenge when proving convergence of Newton-like methods near singular points is ensuring that the iterates remain well-defined. In [37] we introduced a novel safeguarding scheme called γ-safeguarding, defined in Algorithm 7.3 to deal with this problem. The result was a convergence proof for γ-safeguarded AANewton, and it was observed numerically to perform better or no worse compared to standard AANewton, particularly when applied to nonsingular problems.

ALGORITHM 7.3. $\lambda = $ **γ-safeguarding(**$\gamma^{k+1}, w_k, w_{k+1}, r$**).**

Compute $\eta_{k+1} = \|w_{k+1}\|/\|w_k\|$
Set $\beta_{k+1} = r\eta_{k+1}$
if $\gamma^{k+1} = 0$ **or** $\gamma^{k+1} \ge 1$ **then**
    Set $\lambda = 0$
else if $|\gamma^{k+1}|/|1 - \gamma^{k+1}| > \beta_{k+1}$ **then**
    Set $\lambda = \dfrac{\beta_{k+1}}{\gamma^{k+1}\left(\beta_{k+1} + \mathrm{sign}(\gamma^{k+1})\right)}$
else
    Set $\lambda = 1$
end if

[4]That is, Q_{k+1} denotes a term such that $\|Q_{k+1}\| \le |1 - \gamma_{k+1}|\, C\, \|e_k\| \max\{\|e_{k-1}\|, \|e_{k-2}\|\} + |\gamma_{k+1}|\, C\, \|e_{k-1}\| \max\{\|e_{k-2}\|, \|e_{k-3}\|\} + q_{k-1}^k$.

Then γ-Safeguarded AANewton is given by scaling γ^{k+1} in Algorithm 7.2 by $\lambda \in [0,1]$, as computed by Algorithm 7.3. Scaling γ closer to zero scales down the AA correction, moving the update closer to a regular Newton update.

ALGORITHM 7.4. γ-Safeguarded AANewton (γNA(r)).

Choose initial iterate x_0 and $r \in (0,1)$
1: Set $w_1 = -f'(x_0)^{-1} f(x_0)$, and set $x_1 = x_0 + w_1$
2: **for** k=1,2,... **do**
3: Compute $w_{k+1} = -f'(x_k)^{-1} f(x_k)$
4: Compute $\gamma^{k+1} = (w_{k+1} - w_k)^T w_{k+1} / \|w_{k+1} - w_k\|_2^2$
5: Compute $\lambda = \gamma$-safeguarding$(\gamma^{k+1}, w_k, w_{k+1}, r)$
6: Update $x_{k+1} = x_k + w_{k+1} - \lambda\gamma^{k+1}(x_k - x_{k-1} + w_{k+1} - w_k)$
7: **end for**

When $r \approx 0$, the γNA(r) iterates will be heavily scaled towards the latest Newton step, and when $r \approx 1$, the γNA(r) iterates will behave more like standard AANewton. The idea behind γ-safeguarding is to take advantage of the particular convergence behavior of Newton's method near singular points, described in detail in [38].

Next we have an adaptive version of γNA(r) that enjoys guaranteed local convergence when applied to singular problems, but can also detect nonsingular problems automatically, at no additional computational cost, and "turn off" AA in response. This leads to local quadratic convergence if the problem is nonsingular.

ALGORITHM 7.5. λ = Adaptive γ-safeguarding($\gamma^{k+1}, w_k, w_{k+1}, r$).

Compute $\eta_{k+1} = \|w_{k+1}\| / \|w_k\|$
Compute $r_{k+1} = \min\{\eta_{k+1}, r\}$
Set $\beta_{k+1} = r_{k+1}\eta_{k+1}$
if $\gamma^{k+1} = 0$ **or** $\gamma^{k+1} \geq 1$ **then**
Set $\lambda = 0$
else if $|\gamma^{k+1}|/|1-\gamma^{k+1}| > \beta_{k+1}$ **then**
Set $\lambda = \dfrac{\beta_{k+1}}{\gamma^{k+1}\left(\beta_{k+1} + \mathrm{sign}(\gamma^{k+1})\right)}$
else
Set $\lambda = 1$
end if

Adaptive γ-Safeguarded AANewton (γNAA(r)) is given by replacing the call to Algorithm 7.3 with a call to Algorithm 7.5 in line 5 of Algorithm 7.4. The adaptivity in Adaptive γ-safeguarding is the replacement of the constant user-chosen scalar parameter r with $r_{k+1} < r$, where r_{k+1} adapts to the problem based on the computed ratio of residuals given by η_{k+1}.

The formulation of r_{k+1} is based on the satisfaction of three criteria given precisely in [38]. Essentially, if the iterates are at risk for leaving the domain of invertibility of f, then the AA-correction should be small; if the problem is nonsingular, the AA-correction should be scaled to zero; and if the iteration is in the regime of slow linear convergence for Newton's method, the correction step should be scaled closer to standard AANewton.

Local convergence of γNAA(r) follows from Theorem 6.1 in [37]. In particular, let λ_{k+1} be the value of λ computed by Algorithm 7.5 at step k, $\theta_{k+1}^{\lambda} = \|w_{k+1} - \lambda_{k+1}\gamma^{k+1}(w_{k+1} - w_k)\|/\|w_{k+1}\|$, $x_{k+1}^{NA} := x_k + w_{k+1} - \lambda_{k+1}\gamma^{k+1}(x_k - x_{k-1} + w_{k+1} - w_k)$, and $\sigma_k := \|P_R e_k\|/\|P_N e_k\|$. Then we have the following theorem.

Theorem 7.18. *Let* $\dim N = 1$, *and let* $\hat{D}$ *be invertible as a map on* N. *Let* $W_k := B_{\|e_k\|}(x^*) \cap \{x : \|P_R(x - x^*)\| < \sigma_k \|P_N(x - x^*)\|\}$. *If* x_0 *is chosen so that* $\sigma_0 < \hat{\sigma}$ *and* $\|e_0\| < \hat{\rho}$, *for sufficiently small* $\hat{\sigma}$ *and* $\hat{\rho}$, $x_1 = x_0 + w_1$, *and* $x_{k+1} = x_{k+1}^{NA}$ *for* $k \geq 1$, *then* $W_{k+1} \subset W_0$ *for all* $k \geq 0$ *and* $x_k \to x^*$. *That is,* $\{x_k\}$ *remains well-defined and converges to* x^*. *Furthermore, there exist constants* $C > 0$ *and* $\kappa \in (1/2, 1)$ *such that*

$$\|P_R e_{k+1}\| \leq C \max\{|1 - \lambda_{k+1}\gamma^{k+1}|\, \|e_k\|^2, |\lambda_{k+1}\gamma^{k+1}|\, \|e_{k-1}\|^2\}, \tag{7.2.1}$$

$$\|P_N e_{k+1}\| \leq \kappa \theta_{k+1}^{\lambda} \|P_N e_k\|, \tag{7.2.2}$$

for all $k \geq 1$.

Remark 7.19. *Notice this shows the decomposition of the error into two terms, one of which* (7.2.1) *converges with higher order and the other* (7.2.2) *has an improved linear convergence rate. The detailed analysis of [37], summarized in Section 7.1, shows that when* $P_N e_{k+1}$ *dominates, the iterates are well into the region where the Newton iterates remain well-defined, but Newton converges slowly and hence AA improves the rate. When* $P_R e_{k+1}$ *dominates, the iterates are near the singular regime. The heuristic behind the safeguarding is to keep from reducing* $P_N e_{k+1}$ *too fast in comparison to* $P_R e_{k+1}$ *and hence to stay well defined.*

Under the assumptions of Theorem 7.18, Griewank's theorem [67, Theorem 6.1] says that Newton's method almost surely leads into the domain of convergence $\hat{W}$ provided x_0 is sufficiently close to x^*. This effectively means that if the sequence x_k generated by γNAA(r) approaches x^*, we will have almost sure convergence eventually, and this convergence will be faster than Newton. The precise improvement is determined by the asymptotic behavior of θ_{k+1}^{λ}. Globalization techniques such as linesearch methods may be used to bring the γNAA(r) iterates closer to x^*. In particular, γNAA(r) with an Armijo linesearch was shown to be effective in [37].

The main result is Theorem 7.21 below which bounds $\|x_{k+1}^{\text{NA}} - x_{k+1}^{\text{Newt}}\|$ locally, where $x_{k+1}^{\text{Newt}} := x_k + w_{k+1}$, and $x_{k+1}^{\text{NA}} := x_{k+1}^{\text{Newt}} - \lambda_{k+1}\gamma^{k+1}(x_{k+1}^{\text{Newt}} - x_k^{\text{Newt}})$. This notation is introduced to emphasize the Newton and AANewton iterates in the comparison. We will also define $e_k^{\text{Newt}} := x_k^{\text{Newt}} - x^* = x_{k-1} + w_k - x^*$. The following lemma will be used in the proof of Theorem 7.21. Lemma 7.20 bounds $|\lambda_{k+1}\gamma^{k+1}|$, the scaled γ^{k+1} returned by γNAA(r) at iteration k, in terms of η_{k+1} and $r_{k+1} = \min\{\eta_{k+1}, r\}$.

Lemma 7.20. *Let* $\eta_{k+1} = \|w_{k+1}\|/\|w_k\|$, *and let* $r \in (0, 1)$. *Define* $r_{k+1} := \min\{\eta_{k+1}, r\}$ *and* $\beta_{k+1} := r_{k+1}\eta_{k+1}$ *as in Algorithm 7.5. Let* λ_{k+1} *be the value computed by Algorithm 7.5 at iteration* k. *If* $\eta_{k+1} < 1$, *then* $\lambda_{k+1}\gamma^{k+1}$ *returned by Algorithm 7.5 satisfies*

$$|\lambda_{k+1}\gamma^{k+1}| \leq \frac{\beta_{k+1}}{1 - \beta_{k+1}} \tag{7.2.3}$$

when $\lambda_{k+1} = 1$ *and*

$$|\lambda_{k+1}\gamma^{k+1}| = \frac{\beta_{k+1}}{1 + \operatorname{sign}(\gamma^{k+1})\beta_{k+1}} \tag{7.2.4}$$

when $\lambda_{k+1} < 1$.

With Lemma 7.20, we can bound $\|x^{\text{NA}}_{k+1} - x^{\text{Newt}}_{k+1}\|$ in terms of $\|e_k\|$, $\|e_{k-1}\|$, and η_{k+1}.

Theorem 7.21. *If $f'(x^*)$ is nonsingular, then there exist a $\rho > 0$ and a constant C depending only on f such that for x_k and x_{k-1} in $B_\rho(x^*)$ and $\eta_{k+1} < 1$,*

$$\|x^{NA}_{k+1} - x^{Newt}_{k+1}\| \leq C\left(\frac{\beta_{k+1}}{1-\beta_{k+1}}\right)\max\{\|e_k\|^2, \|e_{k-1}\|^2\} \tag{7.2.5}$$

when $\lambda_{k+1} = 1$ and

$$\|x^{NA}_{k+1} - x^{Newt}_{k+1}\| \leq C\left(\frac{\beta_{k+1}}{1+\text{sign}(\gamma^{k+1})\beta_{k+1}}\right)\max\{\|e_k\|^2, \|e_{k-1}\|^2\} \tag{7.2.6}$$

when $\lambda_{k+1} < 1$.

Finally, Corollary 7.22 proves that γNAA(r) can recover local quadratic convergence from NA when applied to nonsingular problems.

Corollary 7.22. *If $f'(x^*)$ is nonsingular, $\|e_k\| < \|e_{k-1}\|$, and $\eta_{k+1} < \hat{r}$, then there exist a $\rho > 0$ and constants C_1 and C_2 depending only on f such that*

$$\|e_{k+1}\| \leq \left(\frac{C_1}{1-\hat{r}^2} + C_2\right)\|e_k\|^2 \tag{7.2.7}$$

for $x_k, x_{k-1} \in B_\rho(x^*)$.

7.3 ▪ Numerics

In this section, we demonstrate the effectiveness of NA and γNAA(r) near bifurcation points by applying these algorithms to the parameter-dependent problem of Navier-Stokes flow in a channel (see Section 3.3) defined as

$$\begin{cases} -\nu\Delta u + u\cdot\nabla u + \nabla p = 0, & u = u_{\text{in}},\ \Gamma_{\text{in}}, \\ & u = 0,\ \Gamma_{\text{wall}}, \\ \nabla\cdot u = 0, & -pn + (\nu\nabla u)n = 0,\ \Gamma_{\text{out}}. \end{cases} \tag{7.3.1}$$

Additional examples on Rayleigh-Bénard flow are shown in [38].

Here, u denotes the fluid velocity, p the pressure, and n the outward normal. In model (7.3.1), ν denotes the viscosity parameter. We use (P_2, P_1) Taylor-Hood elements [23, p. 164]. The channel, shown in Figure 7.1, is arranged such that the left-most boundary lies at $x = 0$, the right-most boundary lies at $x = 50$, and the boundary components are given by $\Gamma_{\text{in}} = \{0\} \times [2.5, 5]$, $\Gamma_{\text{out}} = \{50\}\times[0, 7.5]$, and $\Gamma_{\text{wall}} = [0, 10]\times(\{2.5\}\cup\{5\})\cup\{10\}\times([0, 2.5]\cup[5, 7.5])\cup[10, 40]\times(\{0\}\cup\{7.5\})$. The mesh used is shown in Figure 7.1.

For model (7.3.1), it is known [85] that there exists a critical viscosity $\nu^* \in (0.9, 1)$ at which a bifurcation occurs. For $\nu > \nu^*$, the stable velocity solution is symmetric about the center horizontal ($y = 3.75$) as seen in the top plot of Figure 7.2. For $\nu < \nu^*$, there is still a symmetric solution, but it is unstable. Stability is inherited by two asymmetric solutions seen in the bottom two plots of Figure 7.2.

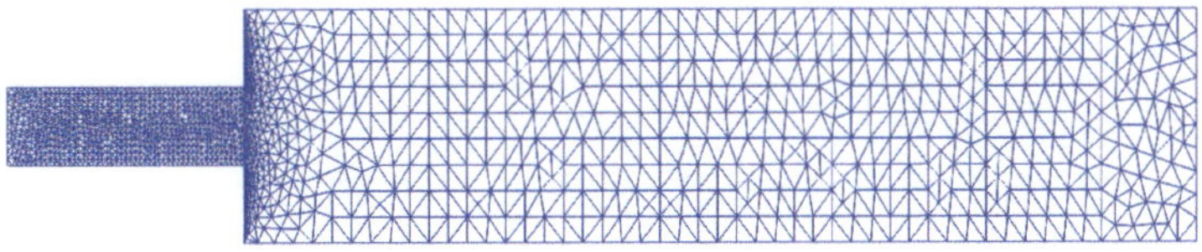

Figure 7.1. *Mesh used for flow in a channel. [38], distributed under CC BY 4.0 DEED License (https://creativecommons.org/licenses/by/4.0/).*

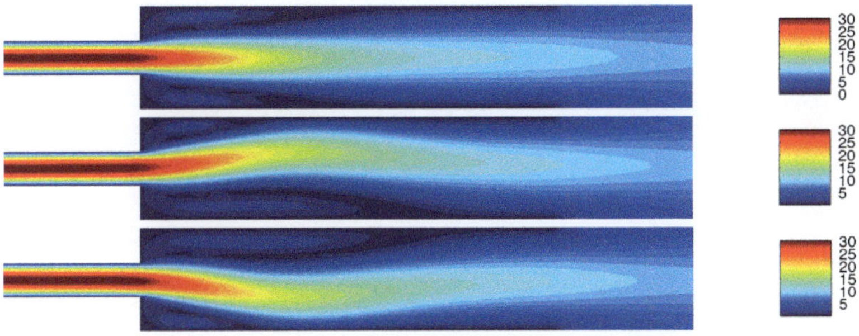

Figure 7.2. *Solutions to channel flow problem* (7.3.1) *for different* ν. *Top: Representative symmetric solution for* $\nu = 1.0$. *Middle: Representative asymmetric solution with positive vertical velocity upon exiting the narrow channel for* $\nu = 0.9$. *Bottom: Representative asymmetric solution with negative vertical velocity upon exiting the narrow channel for* $\nu = 0.9$. *[38], distributed under CC BY 4.0 DEED License (https://creativecommons.org/licenses/by/4.0/).*

In the results shown, we take the zero vector as our initial iterate. The following experiments demonstrate an *asymptotic safeguarding* strategy in which we run AANewton until the residual is smaller than some user-chosen threshold, and we use γNAA(r) for all subsequent iterates. Hence the solve will behave like AANewton until the last few iterations when γNAA(r) is applied. In all experiments we take $r \in (0, 1)$ since this is the range in which local convergence of γNAA(r) is guaranteed by Theorem 7.18. This strategy is further motivated by the work of Pták [95], which shows that within Newton's basin of convergence, even for nonsingular problems, we can expect to see linear convergence first before the asymptotically quadratic convergence takes over. Hence it makes sense to start the iteration by accelerating the linear convergence.

Under the assumptions of Theorem 7.18 or Theorem 7.21, γNAA(r) is guaranteed to converge locally. This motivates our asymptotic safeguarding, in which the iteration starts with AANewton and switches to γNAA(r) once the residual is smaller than a set threshold. This allows the full advantages of AANewton in the preasymptotic regime and ensures fast quadratic convergence for nonsingular problems. In practice, this means fast local convergence provided AANewton reaches the domain of convergence. We activate γNAA(r) when $\|w_{k+1}\| < 10^{-1}$ in the experiments below. When $\|w_{k+1}\| > 10^{-1}$, we run AANewton.

The results are shown below in Figure 7.3. We see that when applied asymptotically, γNAA(r) is not overly sensitive to the choice of parameter r. We also see convergence essentially in agreement with Newton where Newton converges quadratically for $\nu = 0.96$, but fast convergence when Newton has a long preasymptotic regime with $\nu = 0.94$ and $\nu = 0.92$.

Additional numerical results, including the use of γNAA(r) throughout the entire iteration, and results for AANewton(m) with $m > 1$ are shown in [38].

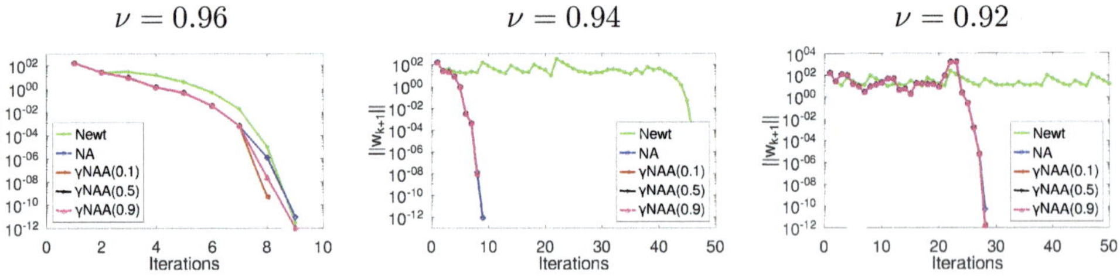

Figure 7.3. *Comparison of γNAA(r) with different values of r applied asymptotically with Newton and NA applied with $\nu = 0.96$ (left), $\nu = 0.94$ (center), and $\nu = 0.92$ (right). Newt is Algorithm 7.1 (Newton), NA is Algorithm 7.2 (AANewton), and γNAA(r) is Algorithm 7.4 with line 5 replaced by a call to Algorithm 7.5. [38], distributed under CC BY 4.0 DEED License (https://creativecommons.org/licenses/by/4.0/).*

Chapter 8
Anderson acceleration's effect on convergence order

My psychiatrist told me I was crazy. I said I want a second opinion. He said "okay, you're ugly too."
—Rodney Dangerfield

In this chapter, which summarizes the main results of [99], we consider the application of AA to iterations which do not converge linearly. The main result from this chapter can be very easily stated: under some assumptions of the fixed-point operator, methods of order r are transformed by AA with $m = 1$ to have order (at least) $\frac{r+1}{2}$. For $m > 1$, the somewhat more complicated result is shown in Theorem 8.8. These results suggest (but do not prove) that AA reduces the asymptotic order of convergence for superlinearly converging methods. Numerical tests from [99] show that for several methods these bounds are sharp and the order is indeed changed to $\frac{r+1}{2}$ by AA; however, an exception is AANewton, for which we observe rate 1.618 (while this theory predicts order at least 1.5). For sublinearly converging methods where locally $r < 1$ (even though asymptotically $r \to 1$), AA will improve convergence.

In this chapter, we refer to the convergence order r as follows. Suppose for k sufficiently large, the iterates satisfy for $\rho > 0$,

$$\|x_{k+1} - x^*\| \leq \rho \|x_k - x^*\|^r. \tag{8.0.1}$$

If $r > 1$, the order of convergence is called superlinear, and if $r = 1$ and $\rho < 1$, the convergence is called linear [10, 63].

There are (at least) two ways to describe a sublinearly convergent iteration, one from convergence rate ρ in (8.0.1), whereas the other one is from convergence order r. If the iteration satisfies

$$\|x_{k+1} - x^*\| \leq \rho_k \|x_k - x^*\|,$$

and $\rho_k < 1$ but $\rho_k \to 1$ as $k \to \infty$, then the convergence is called sublinear [63]. In this chapter, we focus on how AA affects the sublinear convergent iteration from the convergence order aspect, where the sublinearly convergent iteration is defined by a converging sequence satisfying

$$\|x_{k+1} - x^*\| \leq L_k \|x_k - x^*\|^{r_k},$$

with $r_k \leq 1$, $L_k \leq 1$, and $r_k \to 1$ as $k \to \infty$. Herein, we consider a single step (i.e., local) analysis of convergence order, and in a local setting this is a more precise description of a sublinearly converging method. Hence, for a single step, we can discuss the local convergence order r to satisfy (8.0.1) for any $r > 0$. If $r \geq 1$, then the local order agrees with the order.

Assumption 8.1. *Let $g : D \subset X \to X$, with X a Hilbert space and $\|\cdot\|$ the associated norm, be a fixed-point iteration on a bounded convex set D containing x^* and satisfying the following:*

1. *One of the following must hold in D:*

   (a) *There are constants $r > 1$ and $L_0 > 0$ with*
   $$\|g(x+h) - g(x)\| \leq L_0 \|h\|^r.$$

   (b) *g is Lipschitz continuous: $\|g(x+h) - g(x)\| \leq C_0\|h\|$, and there exist a continuous function $r : [0,\infty) \to (0,1]$ satisfying $r(0) = 1$ and a constant $L_0 \leq 1$ satisfying*
   $$\|g(x+h) - g(x)\| \leq L_0 \|h\|^{r(\|x-x^*\|)}.$$

2. *g is Fréchet differentiable, and its derivative $g' : D \times D \to X$ satisfies $\|g'(x;s)\| \leq C_1 \|x\| \|s\|$.*

3. *g' is Lipschitz continuous: $\|g'(x+h;s) - g'(x;s)\| \leq L_1 \|h\| \|s\|$.*

where the constants $C_1, C_0, L_1, L_0 > 0$ are independent of $x, h, s \in X$.

Remark 8.2. *Assumption 8.1 implies the assumptions for the linear case, i.e., Assumptions 4.1 and 4.2, and in particular Lipschitz continuity.*

Remark 8.3. *Part 1.b of Assumption 8.1 has r as a function depending on the distance from a point x to the fixed point x^*. Since our analysis to follow is for a single step of a fixed-point method, we will consider a constant r to be the local sublinear convergence order.*

Since g is Fréchet differentiable, $\|g(x+h) - g(x) - g'(x;h)\|$ is a higher-order term with respect to $\|h\|$ as $\|h\| \to 0$. The next lemma shows the standard result that it is second order in $\|h\|$ as $\|h\| \to 0$ under Assumption 8.1. This result will be used in the analysis below.

Lemma 8.4. *Under Assumption 8.1, for any $x, x+h \in D$ the following inequality is satisfied:*
$$\|g(x+h) - g(x) - g'(x;h)\| \leq \frac{L_1}{2} \|h\|^2. \tag{8.0.2}$$

Proof. By the integral mean value theorem, for any $x, x+h \in D$ we have
$$g(x+h) - g(x) = \int_0^1 g'(x+th;h)\, dt.$$

Taking norms on both sides gives
$$\begin{aligned}
\|g(x+h) - g(x) - g'(x;h)\| &= \left\| \int_0^1 (g'(x+th;h) - g'(x;h))\, dt \right\| \\
&\leq \int_0^1 \|g'(x+th;h) - g'(x;h)\|\, dt \\
&\leq \int_0^1 L_1 \|th\| \|h\|\, dt = \frac{L_1}{2} \|h\|^2,
\end{aligned}$$

due to the Lipschitz continuity of g'. □

With Lemma 8.4, we can now prove results for how AA affects the convergence order of g.

8.1 ▪ Convergence estimates

We first consider $m = 1$ and show the convergence order of g is changed from r to $\frac{r+1}{2}$ for a nonlinear solver satisfying Assumption 8.1.

When $m = 1$, the minimization step in Anderson algorithm (1.2.2) is reduced to

$$\alpha^{k+1} = \text{argmin}_\alpha \|(1-\alpha)w_{k+1} + \alpha w_k\|.$$

For notational simplicity, we define e_j and w_{j+1} as above in Chapter 3 and also

$$\tilde{e}_j = g(x_j) - g(x_{j-1}), w_{j+1}^\alpha = (1-\alpha^{j+1})w_{j+1} + \alpha^{j+1}w_j. \tag{8.1.1}$$

Also, from the definition of θ_j and (1.2.2) we have the following relation (see Chapter 3):

$$|\alpha^{j+1}|\|w_{j+1} - w_j\| = \sqrt{1-\theta_j^2}\|w_{j+1}\|. \tag{8.1.2}$$

This leads to two important inequalities that will allow us to bound the difference between successive iterations by the residual w_j.

Lemma 8.5. *The following inequalities hold for any positive integer j:*

$$\|e_{j-1}\| \le \frac{1}{\sigma|\alpha^j|}\sqrt{1-\theta_{j-1}^2}\|w_j\|, \tag{8.1.3}$$

$$\|e_j\| \le \left(\theta_{j-1} + \frac{1}{\sigma}\sqrt{1+\theta_{j-1}^2}\right)\|w_j\|. \tag{8.1.4}$$

Equation (8.1.3) can be derived directly from the definition of σ and (8.1.2), while (8.1.4) is obtained from

$$\|e_j\| = \|w_j^\alpha - \alpha^j e_{j-1}\| \le \theta_{j-1}\|w_j\| + |\alpha^j|\|e_{j-1}\|,$$

thanks to $w_{j+1}^\alpha = e_{j+1} + \alpha^{j+1}e_j$ and the triangle inequality.

Now we give the result for m=1.

Theorem 8.6 (Convergence order for $m = 1$). *Consider AA applied to the fixed-point iteration defined by $x_{k+1} = g(x_k)$ satisfying Assumption 8.1, with the previous two iterates sufficiently close to the fixed point. We have the one-step bound for the residual w_{k+1}:*

$$\|w_{k+1}\| \le C(\|w_k\|^{\frac{r+1}{2}} + \|w_k\|^2), \tag{8.1.5}$$

where C is a positive constant depending on $\alpha^k, C_1, L_0, L_1, \theta_{k-1}, \sigma, x^*$.

Remark 8.7. *Theorem 8.6 suggests that AA with $m = 1$ reduces the convergence order of superlinearly converging methods but increases the (local) order for sublinearly converging methods. This result is also consistent with what is already known for linearly convergent methods, i.e., AA does not affect the order of linearly converging methods but improves the rate.*

Proof. From the polarization identity, we have

$$\begin{aligned}
&2(w_{k+1}, g'(x_{k-1}; e_k) + \alpha^k g'(x_{k-2}; e_{k-1})) \\
&= \|w_{k+1}\|^2 + \|g'(x_{k-1}; e_k) + \alpha^k g'(x_{k-2}; e_{k-1})\|^2 \\
&- \|w_{k+1} - g'(x_{k-1}; e_k) - \alpha^k g'(x_{k-2}; e_{k-1})\|^2,
\end{aligned}$$

which immediately leads to

$$\begin{aligned}\|w_{k+1}\|^2 \le \|w_{k+1} - g'(x_{k-1}; e_k) - \alpha^k g'(x_{k-2}; e_{k-1})\|^2 \\ + 2(w_{k+1}, g'(x_{k-1}; e_k) + \alpha^k g'(x_{k-2}; e_{k-1})),\end{aligned}$$

by dropping the nonnegative term $\|g'(x_{k-1}; e_k) + \alpha^k g'(x_{k-2}; e_{k-1})\|^2$. Now we bound the right-hand side term by term. To keep the analysis clean, we denote

$$\psi_j = g(x_j) - g(x_{j-1}) - g'(x_{j-1}; e_j).$$

From the triangle inequality, Lemma 8.4, and inequalities (8.1.3)-(8.1.4) we obtain

$$\begin{aligned}\|w_{k+1} - g'(x_{k-1}; e_k) - \alpha^k g'(x_{k-2}; e_{k-1})\|^2 &= \|\psi_k + \alpha^k \psi_{k-1}\|^2 \\ &\le 2\|\psi_k\|^2 + 2|\alpha^k|^2 \|\psi_{k-1}\|^2 \\ &\le \frac{L_1^2}{2}\|e_k\|^4 + |\alpha^k|^2 \frac{L_1^2}{2}\|e_{k-1}\|^4 \\ &\le C\|w_k\|^4,\end{aligned}$$

where C is a constant depending on $\alpha^k, L_1, \theta_{k-1}, \sigma$. Utilizing Assumption 8.1, that g' is bounded, we have

$$\begin{aligned}\|g'(x_{k-1}; e_k) + \alpha^k g'(x_{k-2}; e_{k-1})\| &\le \|g'(x_{k-1}; e_k)\| + |\alpha^k| \|g'(x_{k-2}; e_{k-1})\| \\ &\le C_1\|x_{k-1}\|\|e_k\| + |\alpha^k| C_1 \|x_{k-2}\|\|e_{k-1}\| \le C\|w_k\|,\end{aligned}$$

since the previous two iterates are assumed close to the fixed point.

On the other hand, from the Cauchy-Schwarz inequality and the identity $w_{k+1} = \tilde{e}_k + \alpha^k \tilde{e}_{k-1}$, we get

$$\begin{aligned}&(w_{k+1}, g'(x_{k-1}; e_k) + \alpha_k g'(x_{k-2}; e_{k-1})) \\ &\quad= (\tilde{e}_k + \alpha^k \tilde{e}_{k-1}, g'(x_{k-1}; e_k) + \alpha_k g'(x_{k-2}; e_{k-1})) \\ &\quad\le \|\tilde{e}_k + \alpha^k \tilde{e}_{k-1}\| \|g'(x_{k-1}; e_k) + \alpha_k g'(x_{k-2}; e_{k-1})\| \\ &\quad\le \left(L_0\|e_k\|^r + |\alpha^k| L_0 \|e_{k-1}\|^r\right) C\|w_k\| \le C\|w_k\|^{1+r},\end{aligned}$$

thanks to Assumption 8.1 for the convergence order of g and (8.1.3)-(8.1.4). Combining the inequalities above, we get

$$\begin{aligned}\|w_{k+1}\|^2 &\le 2(w_{k+1}, g'(x_{k-1}; e_k) + \alpha_k g'(x_{k-2}; e_{k-1})) \\ &\quad + \|w_{k+1} - g'(x_{k-1}; e_k) - \alpha_k g'(x_{k-2}; e_{k-1})\|^2 \\ &\le C\left(\|w_k\|^{r+1} + \|w_k\|^4\right).\end{aligned}$$

Taking square-roots on both sides finishes the proof. □

The analysis for general m is more complicated, and we state the result next and refer the reader to [99] for the proof.

Theorem 8.8 (Convergence order for general $m \ge 1$). *Assuming $\alpha_k^{k-m_k} \neq 0$, we have one-step bound of residual w_{k+1} for general depth $m \ge 1$ AA:*

$$\|w_{k+1}\| \le C\left((\|w_k\| + \|w_{k-1}\| + \cdots + \|w_{k-m_k}\|)^{\frac{r+1}{2}} + (\|w_k\| + \|w_{k-1}\| + \cdots + \|w_{k-m_k}\|)^2\right), \tag{8.1.6}$$

where C is a constant depending on problem data.

8.2 ▪ Numerical experiments

In the paper [99], several simple numerical tests are done to test the predicted orders when AA is applied to various solvers: tests show that AA reduces Chebyshev's cubic convergence to quadratic and secant's order of 1.618 to about 1.3, but AANewton converges with a rate about 1.6 (i.e., slightly better than the theory's prediction of 1.5). Also in [99] is a simple example of AA improving sublinear convergence with the improvement of r to $\frac{r+1}{2}$ locally observable. We refer the interested reader to [99] for this illustration of predicted orders.

We show in Figure 8.1 results from a numerical test of AA used for the regularized Bingham model of viscoplastic flow. The Picard iteration for this system is known from numerical tests in [8, 90] to converge sublinearly, and here we show that the convergence order is increased by AA. We consider regularized Bingham and the associated Picard iteration from Section 3.7, using the 2D driven cavity from (4.5.5), with $\tau_s = 5$, $f = 0$ and $\epsilon = 10^{-7}$, $u_0 = 0$, and compute until the L^2 velocity residual falls below 10^{-8}. We use varying m, but no relaxation, since as discussed in [90] relaxation will slow the iteration down for these parameter choices. Convergence plots for varying m are shown in Figure 8.1. The Picard iteration appears to converge sublinearly (and certainly locally sublinear in the early iterations). From the plot data we calculate the local convergence order begins near 0.8 and then slowly increases to 1 once the residual is small. The AAPicard convergence plots are also sublinear (you have to look closely, but the curves do bend), and begin with convergence order calculated to be approximately 0.9 and then increase to near 1, with little difference in convergence for any $m > 0$. These results are consistent with the theory for changing local order r to $\frac{r+1}{2}$.

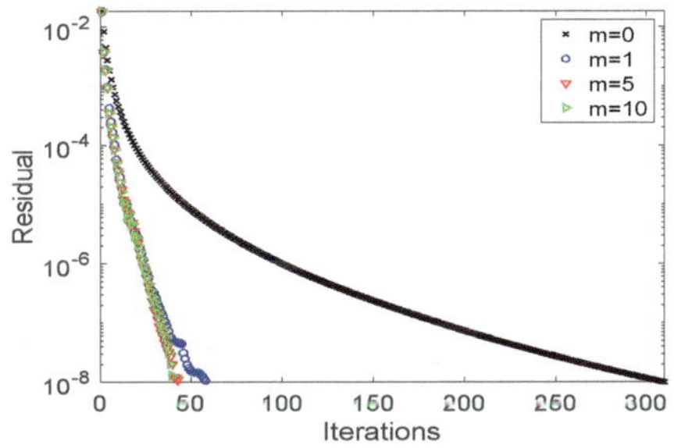

Figure 8.1. *Shown above are convergence plots for the regularized Bingham tests with varying m. Reprinted from [99], with permission from Springer Nature.*

8.3 ▪ Convergence order of AANewton

We now look a bit deeper into the asymptotic convergence order of AANewton, since the theory above predicts an order of (at least) 1.5 but numerical tests in [99] show an order above 1.6. While Chapter 7 reveals how AA can be used to improve Newton in certain settings, in the case of asymptotic quadratic convergence it is well known that AA will decelerate Newton's convergence [50, 92, 99]. The analysis below, which applies results from Chapter 3, gives a different analysis of AANewton's order which leads to a more precise estimate that agrees with numerical tests in the literature.

Denote $g(x) = x - f'(x)^{-1}f(x)$ for a given function $f : X \to X$ for which we seek a root x^*, and $f'(x)^{-1}$ is defined appropriately. Assume that $f'(x^*) \neq 0$, i.e., x^* is not a multiple root, and that g is Lipschitz continuously differentiable. The multiple root case for AANewton is discussed in [87], and it is found that in this case Newton converges linearly, and so the convergence results in Chapter 4 apply directly with θ_k improving the linear convergence rate. We also assume that the initial guess of the algorithm is sufficiently good, which for asymptotic analysis is equivalent to assuming the last m iterates are close to the root. Note that under this last assumption, relaxation would not be used (in the relaxed case, convergence is linear and AA will improve the linear convergence rate).

The key observation for determining AANewton's asymptotic convergence order is to notice that Newton is a single step fixed-point iteration with $g'(x^*) = 0$, and so the convergence results of Theorem 4.5 apply, but now with $\kappa_g = 0$. Hence from Theorem 4.5 we obtain that the AANewton residuals satisfy

$$\|w_{k+1}\| \leq \theta_k(1-\beta_k)\|w_k\| + C\hat{\kappa}_g \|w_k\| \sum_{n=k-m}^{k} \|w_n\|. \tag{8.3.1}$$

Assuming no relaxation, we obtain the estimate

$$\begin{aligned} \|w_{k+1}\| &\leq C\hat{\kappa}_g \|w_k\| \sum_{n=k-m_{k-1}}^{k} \|w_n\| \\ &= C\hat{\kappa}_g \left(\|w_k\|^2 + \|w_k\|\|w_{k-1}\| + ...\|w_k\|\|w_{k-m}\|\right). \end{aligned}$$

In the asymptotic range the last term on the right-hand side is dominant, and so this reduces to

$$\|w_{k+1}\| \lesssim C\hat{\kappa}_g \|w_k\|\|w_{k-m}\|.$$

From here, following theory similar to secant method convergence-order proofs (see, e.g., [11]), the order will be (at least) the largest positive real root of

$$\alpha^{m+1} - \alpha^m - 1 = 0.$$

These expected convergence orders for AANewton are given in Table 8.1 for varying m. It is no coincidence that AANewton with $m = 1$ has the same order as the secant method, since $\|w_{k+1}\| \leq c\|w_k\|\|w_{k-1}\|$ is a key estimate in the secant method convergence proof from which the convergence order is directly determined to be 1.618 [11].

Table 8.1. *Shown above are expected convergence orders for AANewton with varying m. Reprinted from [99], with permission from Springer Nature.*

m	conv order for AANewton
0	2
1	1.618
2	1.466
3	1.380
4	1.325
14	1.147

Remark 8.9. *While this discussion was restricted to Newton, the ideas could also be considered for a general iteration (with fixed-point function satisfying assumptions of Theorem 4.5) when a very large m is chosen.*

Bibliography

[1] P. Africa, D. Arndt, W. Bangerth, B. Blais, M. Fehling, R. Gassmoller, T. Heister, L. Heltai, M. Kronbichler, M. Maier, P. Munch, M. Schreter-Fleischhacker, J. P. Thiele, B. Turcksin, D. Wells, and V. Yushutin. The deal.II Library, Version 9.6. *Journal of Numerical Mathematics*, 32:369–380, 2024. (Cited on p. 17)

[2] A. C. Aitken. XXV.—on Bernoulli's numerical solution of algebraic equations. *Proceedings of the Royal Society of Edinburgh*, 46:289–305, 1927. (Cited on p. 16)

[3] E. Akervik, L. Brandt, D. Henningson, J. Hoepffner, O. Marxen, and P. Schlatter. Steady solutions of the Navier-Stokes equations by selective damping. *Physics of Fluids*, 18(1-4), 2006. (Cited on p. 53)

[4] M. S. Alnæs, J. Blechta, J. Hake, A. Johansson, B. Kehlet, A. Logg, C. Richardson, J. Ring, M. E. Rognes, and G. N. Wells. The FEniCS project version 1.5. *Archive of Numerical Software*, 3(100), 2015. (Cited on p. 48)

[5] H. An, X. Jia, and H.F. Walker. Anderson acceleration and application to the three-temperature energy equations. *Journal of Computational Physics*, 347:1–19, 2017. (Cited on pp. 2, 62)

[6] D. G. Anderson. Iterative procedures for nonlinear integral equations. *Journal of the Association for Computing Machinery*, 12(4):547–560, 1965. (Cited on pp. 1, 13)

[7] D. G. Anderson. Comments on "Anderson acceleration, mixing and extrapolation". *Numerical Algorithms*, 80:135–234, 2019. (Cited on pp. 6, 13)

[8] A. Aposporidis, E. Haber, M.A. Olshanskii, and A. Veneziani. A mixed formulation of the Bingham fluid flow problem: Analysis and numerical solution. *Computer Methods in Applied Mechanics and Engineering*, 200(29):2434–2446, 2011. (Cited on pp. 27, 79, 80, 99)

[9] D. Arnold and J. Qin. Quadratic velocity/linear pressure Stokes elements. In R. Vichnevetsky, D. Knight, and G. Richter, editors, *Advances in Computer Methods for Partial Differential Equations VII*, pages 28–34. IMACS, 1992. (Cited on p. 21)

[10] U. M. Ascher and C. Greif. *A First Course in Numerical Methods*. SIAM, Philadelphia, 2011. (Cited on pp. 22, 95)

[11] K. Atkinson. *An introduction to numerical analysis*. John Wiley, New York, 1991. (Cited on p. 100)

[12] S. Balay, S. Abhyankar, M. F. Adams, S. Benson, J. Brown, P. Brune, K. Buschelman, E. M. Constantinescu, L. Dalcin, A. Dener, V. Eijkhout, J. Faibussowitsch, W. D. Gropp, V. Hapla, T. Isaac, P. Jolivet, D. Karpeev, D. Kaushik, M. G. Knepley, F. Kong, S. Kruger, D. A. May, L. Curfman McInnes, R. Tran Mills, L. Mitchell, T. Munson, J. E. Roman, K. Rupp, P. Sanan, J. Sarich, B. F. Smith, S. Zampini, H. Zhang, H. Zhang, and J. Zhang. PETSc Web page. https://petsc.org/, 2024. (Cited on p. 17)

[13] W. Bao and Q. Du. Computing the ground state solution of Bose–Einstein condensates by a normalized gradient flow. *SIAM Journal on Scientific Computing*, 25(5):1674–1697, 2004. (Cited on p. 24)

[14] G. Baruch, G. Fibich, and S. Tsynkov. High-order numerical method for the nonlinear Helmholtz equation with material discontinuities in one space dimension. *Journal of Computational Physics*, 227:820–850, 2007. (Cited on p. 21)

[15] G. Baruch, G. Fibich, and S. Tsynkov. A high-order numerical method for the nonlinear Helmholtz equation in multidimensional layered media. *Journal of Computational Physics*, 228:3789–3815, 2009. (Cited on p. 21)

[16] M. Benzi and M. A. Olshanskii. An augmented Lagrangian-based approach to the Oseen problem. *SIAM Journal on Scientific Computing*, 28:2095–2113, 2006. (Cited on p. 28)

[17] W. Bian and X. Chen. Anderson acceleration for nonsmooth fixed point problems. *SIAM Journal on Numerical Analysis*, 60(5):2565–2591, 2022. (Cited on pp. 2, 16)

[18] W. Bian, X. Chen, and C. T. Kelley. Anderson acceleration for a class of nonsmooth fixed-point problems. *SIAM Journal on Scientific Computing*, 43(5):S1–S20, 2021. (Cited on pp. 2, 16)

[19] E. C. Bingham. *Fluidity and Plasticity*. McGraw-Hill, 1922. (Cited on p. 26)

[20] S. Börm and S. Le Borne. $\mathcal{H}$-LU factorization in preconditioners for augmented Lagrangian and grad-div stabilized saddle point systems. *International Journal for Numerical Methods in Fluids*, 68:83–98, 2012. (Cited on p. 28)

[21] J. W. Both, K. Kumar, J. M. Nordbotten, and F. A. Radu. Anderson accelerated fixed-stress splitting schemes for consolidation of unsaturated porous media. *Computers & Mathematics with Applications*, 77(6):1479–1502, 2019. (Cited on p. 2)

[22] N. Boullé, V. Dallas, and P.E. Farrell. Bifurcation analysis of two-dimensional Rayleigh-Bénard convection using deflation. *Physical Review E*, 105, 2022. (Cited on p. 81)

[23] D. Braess. *Finite Elements: Theory, Fast Solvers, and Applications in Solid Mechanics*. Cambridge University Press, 2007. (Cited on p. 92)

[24] S. C. Brenner and L. R. Scott. *The mathematical theory of finite element methods*, volume 15 of *Texts in Applied Mathematics*. Springer, New York, third edition, 2008. (Cited on pp. 19, 20)

[25] C. Brezinski, S. Cipolla, M. Redivo-Zaglia, and Y. Saad. Shanks and Anderson-type acceleration techniques for systems of nonlinear equations. *IMA Journal of Numerical Analysis*, 42:3058–3093, 2022. (Cited on p. 2)

[26] C. Brezinski and M. Redivo-Zaglia. *Extrapolation Methods: Theory and Practice*. North-Holland, Amsterdam, 1991. (Cited on p. 16)

[27] C. Brezinski, M. Redivo-Zaglia, and Y. Saad. Shanks sequence transformations and Anderson acceleration. *SIAM Review*, 60(3):646–669, 2018. (Cited on pp. 1, 16)

[28] C. G. Broyden. A class of methods for solving nonlinear simultaneous equations. *Mathematics of Computation*, 19:577–593, 1965. (Cited on p. 10)

[29] P. R. Brune, M. G. Knepley, B. F. Smith, and X. Tu. Composing scalable nonlinear algebraic solvers. *SIAM Review*, 57(4):535–565, 2015. (Cited on pp. 17, 26, 48)

[30] S. Carl, V. K. Le, and D. Motreanu. *Nonsmooth variational problems and their inequalities: Comparison principles and applications.* Springer monographs in mathematics. New York: Springer Science+Business Media, 2007. (Cited on p. 25)

[31] T. F. Chan. Algorithm 581: An improved algorithm for computing the singular value decomposition [f1]. *ACM Transactions on Mathematical Software*, 8(1):84–88, 1982. (Cited on p. 63)

[32] X. Chen and C. T. Kelley. Convergence of the EDIIS algorithm for nonlinear equations. *SIAM Journal on Scientific Computing*, 41(1):A365–A379, 2019. (Cited on pp. 1, 13, 14)

[33] S.-H. Chiu, T.-W. Pan, J. He, A. Guo, and R. Glowinski. A numerical study of the transition to oscillatory flow in 3d lid-driven cubic cavity flows. *arxiv:1604.06926*, 2016. (Cited on p. 53)

[34] M. Chupin, M.-S. Dupuy, G. Legendre, and É. Séré. Convergence analysis of adaptive DIIS algorithms with application to electronic ground state calculations. *ESAIM: Mathematical Modelling and Numerical Analysis*, 55:2785 – 2825, 2021. (Cited on p. 9)

[35] A. Cibik and S. Kaya. A projection-based stabilized finite element method for steady state natural convection problem. *Journal of Mathematical Analysis and Applications*, 381:469–484, 2011. (Cited on pp. 74, 76)

[36] S. Congreve and T. P. Wihler. Iterative Galerkin discretizations for strongly monotone problems. *Journal of Computational and Applied Mathematics*, 311:457–472, 2017. (Cited on p. 25)

[37] M. Dallas and S. Pollock. Newton-Anderson at singular points. *International Journal of Numerical Analysis and Modeling*, 20(5):667–692, 2023. (Cited on pp. 3, 81, 82, 83, 89, 91)

[38] M. Dallas, S. Pollock, and L. Rebholz. Analysis of an adaptive safeguarded Newton-Anderson algorithm of depth one with applications to fluid problems. *Advances in Computational Science and Engineering*, 2(3):246–270, 2024. (Cited on pp. 3, 81, 90, 92, 93, 94)

[39] H. De Sterck and Y. He. On the asymptotic linear convergence speed of Anderson acceleration, Nesterov acceleration, and nonlinear GMRES. *SIAM Journal on Scientific Computing*, 43(5):S21–S46, 2021. (Cited on p. 16)

[40] E. Dean, R. Glowinski, and G. Guidoboni. On the numerical simulation of Bingham visco-plastic flow: Old and new results. *Journal of Non-Newtonian Fluid Mechanics*, 142:36–62, 2007. (Cited on p. 26)

[41] E. J. Dean and R. Glowinski. Operator-splitting methods for the simulation of Bingham visco-plastic flow. *Chinese Annals of Mathematics*, 23(02):187–204, 2002. (Cited on p. 27)

[42] D. W. Decker, H. B. Keller, and C. T. Kelley. Convergence rates for Newton's method at singular points. *SIAM Journal on Numerical Analysis*, 20(2):296–314, 1983. (Cited on pp. 81, 82, 83)

[43] D. W. Decker and C. T. Kelley. Newton's method at singular points. I. *SIAM Journal on Numerical Analysis*, 17(1):66–70, 1980. (Cited on pp. 81, 83)

[44] J. E. Dennis, Jr., and J. J. Moré. Quasi-Newton methods, motivation and theory. *SIAM Review*, 19(1):46–89, 1977. (Cited on p. 11)

[45] P. Deuflhard. *Newton Methods for Nonlinear Problems*. Springer Series in Computational Mathematics. Springer, 2005. (Cited on p. 82)

[46] J. I. Diaz and F. De Thelin. On a nonlinear parabolic problem arising in some models related to turbulent flows. *SIAM Journal on Mathematical Analysis*, 25(4):1085–1111, 1994. (Cited on p. 26)

[47] L. Diening, M. Fornasier, R. Tomasi, and M. Wank. A relaxed Kačanov iteration for the p-Poisson problem. *Numerische Mathematik*, 145:1–34, 2020. (Cited on pp. 26, 48)

[48] H. Elman, D. Silvester, and A. Wathen. *Finite Elements and Fast Iterative Solvers with applications in incompressible fluid dynamics*. Numerical Mathematics and Scientific Computation. Oxford University Press, Oxford, 2014. (Cited on p. 28)

[49] A. Ern and J. L. Guermond. *Theory and Practice of Finite Elements*, volume 159 of *Applied Mathematical Sciences*. Springer-Verlag, New York, 2004. (Cited on p. 20)

[50] C. Evans, S. Pollock, L. Rebholz, and M. Xiao. A proof that Anderson acceleration improves the convergence rate in linearly converging fixed-point methods (but not in those converging quadratically). *SIAM Journal on Numerical Analysis*, 58:788–810, 2020. (Cited on pp. 2, 17, 22, 23, 31, 32, 33, 36, 44, 47, 81, 99)

[51] L. C. Evans. *Partial Differential Equations*. American Mathematical Society, Providence, RI, 2 edition, 2010. (Cited on p. 19)

[52] V. Eyert. A comparative study on methods for convergence acceleration of iterative vector sequences. *Journal of Computational Physics*, 124(2):271–285, 1996. (Cited on pp. 1, 2, 9, 11)

[53] H. Fang and Y. Saad. Two classes of multisecant methods for nonlinear acceleration. *Numerical Linear Algebra with Applications*, 16(3):197–221, 2009. (Cited on pp. 1, 2, 4, 9, 10, 11, 12, 13, 61, 63)

[54] P. E. Farrell, Á. Birkisson, and S.W. Funke. Deflation techniques for finding distinct solutions of nonlinear partial differential equations. *SIAM Journal on Scientific Computing*, 37:A2026–A2045, 2015. (Cited on p. 81)

[55] Y. Feldman and A.Y. Gelfgat. Oscillatory instability of a three-dimensional lid-driven flow in a cube. *Physics of Fluids*, 22(093602):1–9, 2010. (Cited on p. 53)

[56] G. Fibich and S.V. Tsynkov. High-order two-way artificial boundary conditions for nonlinear wave propagation with backscattering. *Journal of Computational Physics*, 171:632–677, 2001. (Cited on p. 21)

[57] D. Forbes, L. Rebholz, and F. Xue. Anderson acceleration of nonlinear solvers for the stationary Gross-Pitaevskii equation. *Advances in Applied Mathematics and Mechanics*, 13:1096–1125, 2021. (Cited on pp. 24, 25, 68)

[58] I. A. Frigaard and C. Nouar. On the usage of viscosity regularisation methods for visco-plastic fluid flow computation. *Journal of Non-Newtonian Fluid Mechanics*, 127(1):1–26, 2005. (Cited on p. 27)

[59] A. Fu, J. Zhang, and S. Boyd. Anderson accelerated Douglas–Rachford splitting. *SIAM Journal on Scientific Computing*, 42(6):A3560–A3583, 2020. (Cited on p. 2)

[60] E. M. Garau, P. Morin, and C. Zuppa. Convergence of an adaptive Kačanov FEM for quasi-linear problems. *Applied Numerical Mathematics*, 61(4):512–529, 2011. (Cited on p. 25)

[61] D. J. Gardner, L. L. LoDestro, and C. S. Woodward. Towards the use of Anderson Acceleration in coupled transport-gyrokinetic turbulence simulations. *CoRR*, abs/2407.03561, 2024. (Cited on p. 17)

[62] D. J. Gardner, D. R. Reynolds, C. S. Woodward, and C. J. Balos. Enabling new flexibility in the SUNDIALS suite of nonlinear and differential/algebraic equation solvers. *ACM Transactions on Mathematical Software (TOMS)*, 2022. (Cited on p. 17)

[63] W. Gautschi. *Numerical Analysis*. Birkhäuser, New York, 2012. (Cited on p. 95)

[64] V. Girault and P.-A. Raviart. *Finite Element Methods for Navier–Stokes Equations: Theory and Algorithms*. Springer-Verlag, 1986. (Cited on pp. 20, 53, 75)

[65] G. H. Golub and C. F. Van Loan. *Matrix Computations (3rd Ed.)*. Johns Hopkins University Press, Baltimore, MD, USA, 1996. (Cited on pp. 12, 13, 32, 42, 65)

[66] A. O. Griewank. *Analysis and Modification of Newton's Method at Singularities*. PhD thesis, Australian National University, 1980. (Cited on p. 82)

[67] A. O. Griewank. Starlike domains of convergence for Newton's method at singularities. *Numerische Mathematik*, 35:95–111, 1980. (Cited on pp. 81, 83, 91)

[68] P. P. Grinevich and M. A. Olshanskii. An iterative method for the Stokes-type problem with variable viscosity. *SIAM Journal on Scientific Computing*, 31(5):3959–3978, 2009. (Cited on p. 27)

[69] J. Guzman and L.R. Scott. The Scott-Vogelius finite elements revisited. *Mathematics of Computation*, 88(316):515–529, 2019. (Cited on pp. 20, 21)

[70] R. Haelterman, J. Degroote, D. Van Heule, and J. Vierendeels. The quasi-Newton least squares method: A new and fast secant method analyzed for linear systems. *SIAM Journal on Numerical Analysis*, 47(3):2347–2368, 2009. (Cited on p. 16)

[71] T. Heister and G. Rapin. Efficient augmented Lagrangian-type preconditioning for the Oseen problem using grad-div stabilization. *International Journal for Numerical Methods in Fluids*, 71:118–134, 2013. (Cited on p. 28)

[72] N. C. Henderson and R. Varadhan. Damped Anderson acceleration with restarts and monotonicity control for accelerating EM and EM-like algorithms. *Journal of Computational and Graphical Statistics*, 28(4):834–846, 2019. (Cited on p. 17)

[73] N. Higham and N. Strabic. Anderson acceleration of the alternating projections method for computing the nearest correlation matrix. *Numerical Algorithms*, 72:1021–1042, 2016. (Cited on p. 2)

[74] V. John. *Finite Element Methods for Incompressible Flow Problems*. Springer, Berlin, 2016. (Cited on p. 53)

[75] V. John, A. Linke, C. Merdon, M. Neilan, and L. G. Rebholz. On the divergence constraint in mixed finite element methods for incompressible flows. *SIAM Review*, 59(3):492–544, 2017. (Cited on pp. 20, 21)

[76] C. T. Kelley. *Solving Nonlinear Equations with Iterative Methods: Solvers and Examples in Julia*. SIAM, Philadelphia, 2022. (Cited on p. 17)

[77] C.T. Kelley. Numerical methods for nonlinear equations. *Acta Numerica*, 27:207–287, 2018. (Cited on pp. 1, 4, 13, 14, 32, 33, 37)

[78] H. Kielhöfer. *Bifurcation Theory: An Introduction with Applications to Partial Differential Equations*. Springer Science+Business Media, 2012. (Cited on p. 81)

[79] L. Landau and E. Lifschitz. *Quantum Mechanics: Non-relativistic Theory*. Pergamon Press, New York, 1977. (Cited on p. 24)

[80] W. Layton. *An Introduction to the Numerical Analysis of Viscous Incompressible Flows*. SIAM, Philadelphia, 2008. (Cited on pp. 19, 20, 23, 74, 76)

[81] P. A. Lott, H. F. Walker, C. S. Woodward, and U. M. Yang. An accelerated Picard method for nonlinear systems related to variably saturated flow. *Advances in Water Resources*, 38:92–101, 2012. (Cited on p. 2)

[82] I. Muga and K. G. van der Zee. Discretization of linear problems in Banach spaces: Residual minimization, nonlinear Petrov–Galerkin, and monotone mixed methods. *SIAM Journal on Numerical Analysis*, 58(6):3406–3426, 2020. (Cited on p. 26)

[83] J. M. Ortega. The Newton-Kantorovich theorem. *The American Mathematical Monthly*, 75(6):658–660, 1968. (Cited on p. 81)

[84] Y. Peng, B. Deng, J. Zhang, F. Geng, W. Qin, and L. Liu. Anderson acceleration for geometry optimization and physics simulation. *ACM Transactions on Graphics*, 42:1–14, 2018. (Cited on p. 2)

[85] F. Pichi. *Reduced order models for parametric bifurcation problems in nonlinear PDEs.* PhD thesis, Scuola Internazionale Supeiore di Studi Avanzati, 2020. (Cited on p. 92)

[86] L. P. Pitaevskii. Vortex lines in an imperfect Bose gas. *Journal of Experimental and Theoretical Physics*, 13:451–454, 1961. (Cited on p. 24)

[87] S. Pollock. Anderson acceleration for degenerate and nondegenerate problems. *Deterministic and Stochastic Optimal Control and Inverse Problems, 1st edition*, pages 1–20, 2021. (Cited on p. 99)

[88] S. Pollock and L. Rebholz. Anderson acceleration for contractive and noncontractive operators. *IMA Journal of Numerical Analysis*, 41(4):2841–2872, 2021. (Cited on pp. 1, 2, 4, 12, 22, 23, 26, 31, 34, 51, 52, 54, 62, 64, 67, 68, 74, 80)

[89] S. Pollock and L. Rebholz. Filtering for Anderson acceleration. *SIAM Journal on Scientific Computing*, 45(4):A1571–A1590, 2023. (Cited on pp. 4, 12, 22, 25, 26, 61, 64, 67, 68, 69, 70)

[90] S. Pollock, L. Rebholz, and D. Vargun. An efficient nonlinear solver and convergence analysis for a viscoplastic flow model. *Numerical Methods for Partial Differential Equations*, 39:3874–3896, 2023. (Cited on pp. 27, 33, 57, 58, 59, 74, 79, 80, 99)

[91] S. Pollock, L. G. Rebholz, and M. Xiao. Anderson-accelerated convergence of Picard iterations for incompressible Navier–Stokes equations. *SIAM Journal on Numerical Analysis*, 57:615– 637, 2019. (Cited on pp. 1, 2, 4, 16, 22, 23, 32, 33, 37, 53, 74, 77)

[92] S. Pollock, L. Rebholz, and M. Xiao. Acceleration of nonlinear solvers for natural convection problems. *Journal of Numerical Mathematics*, 29:1–19, 2021. (Cited on pp. 29, 55, 56, 74, 77, 78, 81, 99)

[93] S. Pollock and H. Schwartz. Benchmarking results for the Newton-Anderson method. *Results in Applied Mathematics*, 8(100095), 2020. (Cited on pp. 81, 84)

[94] F. A. Potra and H. Engler. A characterization of the behavior of the Anderson acceleration on linear problems. *Linear Algebra and Its Applications*, 438(3):1002–1011, 2013. (Cited on p. 16)

[95] V. Pták. Concerning the rate of convergence of Newton's process. *Commentationes Mathematicae Universitatis Carolinae*, 16(4):699–705, 1975. (Cited on p. 93)

[96] P. Pulay. Convergence acceleration of iterative sequences: The case of SCF iteration. *Chemical Physics Letters*, 73:393–398, 1980. (Cited on p. 9)

[97] P. Pulay. Improved SCF convergence acceleration. *J. Comput. Chem.*, 3:556–560, 1982. (Cited on p. 9)

[98] L. Rebholz, D. Vargun, and M. Xiao. Enabling fast convergence of the iterated penalty Picard iteration with $\mathcal{O}(1)$ penalty parameter for incompressible Navier-Stokes via Anderson acceleration. *Computer Methods in Applied Mechanics and Engineering*, 387(114178):1–17, 2021. (Cited on p. 67)

[99] L. Rebholz and M. Xiao. The effect of Anderson acceleration on the convergence order of superlinear and sublinear nonlinear solvers. *Journal of Scientific Computing*, 96(34):1–23, 2023. (Cited on pp. 81, 95, 98, 99, 100)

[100] G.W. Reddien. On Newton's method for singular problems. *SIAM Journal on Numerical Analysis*, 15(5):993–996, 1978. (Cited on p. 81)

[101] G.W. Reddien. Newton's method and high order singularities. *Computers & Mathematics with Applications*, 5(2):79–86, 1979. (Cited on pp. 81, 83)

[102] T. Rohwedder and R. Schneider. An analysis for the DIIS acceleration method used in quantum chemistry calculations. *Journal of Mathematical Chemistry*, 49:1889–1914, 2011. (Cited on p. 9)

[103] Y. Saad. Krylov subspace methods for solving large unsymmetric linear systems. *Mathematics of Computation*, 37:105–126, 1981. (Cited on p. 16)

[104] Y. Saad and M. H. Schultz. GMRES: A generalized minimal residual algorithm for solving nonsymmetric linear systems. *SIAM Journal on Scientific and Statistical Computing*, 7(3):856–869, 1986. (Cited on pp. 14, 15)

[105] H. Schmitt. Numerical simulation of Bingham fluid flow using prox-regularization. *Journal of Optimization Theory and Applications*, 106(3):603–626, 2000. (Cited on p. 27)

[106] R. Schneider, T. Rohwedder, A. Neelov, and J. Blauert. Direct minimization for calculating invariant subspaces in density functional computations of the electronic structure. *Journal of Computational Mathematics*, 27(2/3):360–387, 2009. (Cited on p. 9)

[107] D. Scieur, A. d' Aspremont, and F. Bach. Regularized nonlinear acceleration. In D. Lee, M. Sugiyama, U. Luxburg, I. Guyon, and R. Garnett, editors, *Advances in Neural Information Processing Systems*, volume 29. Curran Associates, Inc., 2016. (Cited on p. 61)

[108] J. Sherman and W. J. Morrison. Adjustment of an inverse matrix corresponding to changes in the elements of a given column or a given row of the original matrix. *The Annals of Mathematical Statistics*, 20:621, 1949. (Cited on p. 10)

[109] W. Shi, S. Song, H. Wu, Y.-C. Hsu, C. Wu, and G. Huang. Regularized Anderson acceleration for off-policy deep reinforcement learning. In H. Wallach, H. Larochelle, A. Beygelzimer, F. d' Alché-Buc, E. Fox, and R. Garnett, editors, *Advances in Neural Information Processing Systems*, volume 32. Curran Associates, Inc., 2019. (Cited on p. 61)

[110] A. Sidi. Minimal polynomial and reduced rank extrapolation methods are related. *Advances in Computational Mathematics*, 43:151–170, 2017. (Cited on p. 16)

[111] D. A. Smith, W. F. Ford, and A. Sidi. Extrapolation methods for vector sequences. *SIAM Review*, 29(2):199–233, 1987. (Cited on p. 16)

[112] P. Stasiak and M.W. Matsen. Efficiency of pseudo-spectral algorithms with Anderson mixing for the SCFT of periodic block-copolymer phases. *The European Physical Journal E*, 34:110:1–9, 2011. (Cited on p. 2)

[113] H. De Sterck. A nonlinear GMRES optimization algorithm for canonical tensor decomposition. *SIAM Journal on Scientific Computing*, 34(3):A1351–A1379, 2012. (Cited on p. 16)

[114] H. De Sterck and Y. He. On the asymptotic linear convergence speed of Anderson acceleration, Nesterov acceleration, and nonlinear GMRES. *SIAM Journal on Scientific Computing*, 43(5):S21–S46, 2021. (Cited on p. 2)

[115] H. De Sterck and Y. He. Linear asymptotic convergence of Anderson acceleration: Fixed-point analysis. *SIAM Journal on Matrix Analysis and Applications*, 43(4):1755–1783, 2022. (Cited on p. 2)

[116] H. De Sterck and Y. He. Anderson acceleration as a Krylov method with application to asymptotic convergence analysis. *Journal of Scientific Computing*, 99:1–30, 2024. (Cited on pp. 2, 61)

[117] R. Temam. Sur l'approximation des solutions des équations de Navier-Stokes. *Comptes Rendus de l'Académie des Sciences Paris, Serie A*, 262:219–221, 1966. (Cited on p. 23)

[118] R. Temam. Sur la stabilité et al convertence de la méthode des pas fractionnaires. *Annali di Matematica Pura ed Applicata*, 79(4):191–379, 1968. (Cited on p. 23)

[119] R. Temam. *Navier-Stokes equations*. Elsevier, North-Holland, 1991. (Cited on p. 23)

[120] A. Toth and C. T. Kelley. Convergence analysis for Anderson acceleration. *SIAM Journal on Numerical Analysis*, 53(2):805–819, 2015. (Cited on pp. 1, 4, 9, 13, 14, 16, 32, 33, 37)

[121] A. Toth, C. T. Kelley, S. Slattery, S. Hamilton, K. Clarno, and R. Pawlowski. Analysis of Anderson acceleration on a simplified neutronics/thermal hydraulics system. *Proceedings of the ANS MC2015 Joint International Conference on Mathematics and Computation (M&C), Supercomputing in Nuclear Applications (SNA) and the Monte Carlo (MC) Method*, ANS MC2015 CD:1–12, 2015. (Cited on p. 2)

[122] H. Uecker. Continuation and bifurcation in nonlinear PDEs – algorithms, applications, and experiments. *Jahresbericht der Deutschen Mathematiker-Vereinigung*, 124:43–80, 2021. (Cited on p. 81)

[123] H. F. Walker and P. Ni. Anderson acceleration for fixed-point iterations. *SIAM Journal on Numerical Analysis*, 49(4):1715–1735, 2011. (Cited on pp. 1, 4, 9, 12, 13, 15, 62)

[124] D. Wang, Y. He, and H. De Sterck. On the asymptotic linear convergence speed of Anderson acceleration applied to ADMM. *Journal of Scientific Computing*, 88:38, 2021. (Cited on pp. 2, 16)

[125] T. Washio and C.W. Oosterlee. Krylov subspace acceleration for nonlinear multigrid schemes. *ETNA. Electronic Transactions on Numerical Analysis*, 6:271–290, 1997. (Cited on p. 16)

[126] M. A. Woodbury. Inverting modified matrices. Technical report, Princeton University, Princeton, NJ, 1950. (Cited on p. 10)

[127] E. Zeidler. *Nonlinear Functional Analysis and its Applications*. Springer, New York, 2013. (Cited on p. 25)

[128] J. Zhang, B. O'Donoghue, and S. Boyd. Globally convergent Type-I Anderson acceleration for nonsmooth fixed-point iterations. *SIAM Journal on Optimization*, 30(4):3170–3197, 2020. (Cited on p. 16)

[129] S. Zhang. A new family of stable mixed finite elements for the 3d Stokes equations. *Mathematics of Computation*, 74(250):543–554, 2005. (Cited on p. 21)

Index